SpringerBriefs in Electrical and Computer Engineering

For further volumes:
http://www.springer.com/series/10059

Mirjana Pavlovic • Bela Balint

Stem Cells and Tissue Engineering

 Springer

Mirjana Pavlovic
Florida Atlantic University
Boca Raton, FL, USA

Bela Balint
Military Medical Academy
Belgrade, Serbia

ISSN 2191-8112   ISSN 2191-8120 (electronic)
ISBN 978-1-4614-5504-2 ISBN 978-1-4614-5505-9 (eBook)
DOI 10.1007/978-1-4614-5505-9
Springer New York Heidelberg Dordrecht London

Library of Congress Control Number: 2012950311

Printed on acid-free paper

Springer is part of Springer Science+Business Media (www.springer.com)

This book is dedicated to our mutual friend and colleague, creative researcher, and dedicated scientist Dr Nevenka Stojanovic, PhD, whom we were lucky to share time in this life with, for a while ...

Authors

Preface

The attractive field of stem cell research raised considerable number of questions suggesting that tremendous piece of work is expected to be done in the near future. It is believed that within the next 6–10 years, stem cell-based applications will be used for the treatment of diabetes mellitus, heart disease, and Parkinson disease. The pluripotency of stem cells from different sources is the subject of controversies and criticism, but as one of the most prominent features of these cells is also envisioned as a powerful therapeutic approach [1–3]. Adult stem cells reside in most mammalian tissues, but the extent to which they contribute to normal homeostasis and repair varies widely.

In the near future, the scientists from this field envision treatments for multiple sclerosis and the damaged nerves of paraplegics. Yet, the stem cell regenerative therapy is not the answer to all diseases. For example, the new treatments are also emerging for Alzheimer's disease, but Alzheimer's strikes look as more likely to be solved by vaccines aimed at beta-amyloid proteins than by stem cell therapy [3]. For the average person who does not have type I diabetes the estimates for heart disease and Parkinson's are more exciting. A stem cell treatment for Parkinson's might also help improve coordination and perhaps even cognitive function among the aged.

Curiously, more researchers are optimistic about the future of treatments based on adult than on embryonic stem cells. The reason for this is relatively clearly explained in this book. Do the researchers see more difficult scientific problems or political obstacles for embryonic stem cells? [4]. Also, there is a specific issue of organ replacement. Some form of stem cells will be the starting point for growing most types of replacement organs. It would be very interesting to know what the stem cell researchers see as the time line for solving stem cell and tissue engineering problems associated with growing replacement organs [5, 6].

Once stem cells can be coaxed to go in and replace lost cells and to grow into replacement organs, human life extension will become a reality. Building and installing replacement parts in humans are much more difficult than in machine. Yet, all of the problems involved in human parts replacement are solvable. Some stem cell and tissue engineering researchers expect many of those problems to be solved in the next 20 years. Let us recapitulate the most critical issues in stem cell transplantation and its use in tissue engineering from ethical, technical, and clinical point of view with the goal of estimation of its efficiency and perspective.

References

1. Hanson C, Hardarson T, Ellerström C, Nordberg M, Caisander, G, Rao M, Hyllner J, Stenevi U (2012) Transplantation of human embryonic stem cells onto a partially wounded human cornea in vitro. Acta Ophtalmol Scan: 1–4
2. Giam LR, Massich MD, Hao L, Wong LS, Mader CC, Mirkin CA (2012) Scanning probe-enabled nanocombinatorics defines the relationship between fibronectin feature size and stem cell fate. PNAS 109(12):4377–4382
3. Singh AM, Dalton S et al (2012) Signaling network crosstalk in human pluripotent cells: a Smad2/3-regulated switch that controls the balance between self-renewal and differentiation p312. Cell Stem Cell 10(3):312–326, 2. doi:10.1016/j.stem.2012.01.014
4. Chhabra A, Lechner JA, Ueno M, Acharya A, Van Handeel B, Wang Y, Iruela-Arispe ML, Tallquist MD, Mikkola HKA (2012) Trophoblast regulates the placental hematopoietic niche through PDGF-B signalling. Dev. Cell 22(3):651–659
5. Kolehmainen K, Willerth SM (2012) Preparation of 3D fibrin scaffolds for stem cell culture applications. J Vis Exp (61):e3641. doi:10.3791/3641
6. Tan CJT, Rahman R, Jaber-Hijazi F, Felix DA, Chen C, Louis EJ, Aboobaker A (2012) Telomere maintenance and telomerase activity are differentially regulated in asexual and sexual worms. PNAS Early Edition: 1–6

Contents

Chapter 1
Short History of Stem Cells Transplantation with Emphasis on Hematological Disorders

The beginning is the most important part of the work.

Plato, Republic

Stem Cell History

Bone Marrow Transplantation

Stem cell history started almost a century ago with administration of bone marrow by mouth to the patients with anemia and leukemia [1]. At that time, in the early 1900s, European scientists realized that all blood cells originate from one particular "stem cell" located within the bone marrow, tissue where the entire hematopoiesis take place. Early studies with animals quickly revealed that the bone marrow was the organ most sensitive to the damaging effects of radiation. Very soon, it became clear that reinfusion of marrow cells could rescue lethally irradiated animals. Mice with defective bone marrow could be restored to health with infusions into the blood stream of marrow taken from other mice [2].

Among the early attempts in humans to do this were several transplants carried out in France following a radiation accident in the late 1950s (Vinca, former Yugoslavia). The first successful bone marrow transplantation was performed at the University of Minnesota in 1968 on a 4-month-old boy suffering from a severe combined immunodeficiency disease ("bubble boy syndrome"). A sibling of the boy had already passed away from the same disease. His body accepted bone marrow extracted from his sister, and his immune system was restored. That boy is now a healthy man—fully employed, living in Connecticut and the proud father of twins.

The beginning of transplantation as a routine therapy started with critical discovery about the human immune system by French medical researcher (1958) Jean Dausset who described the first of many human histocompatibility antigens.

M. Pavlovic and B. Balint, *Stem Cells and Tissue Engineering*,
SpringerBriefs in Electrical and Computer Engineering,
DOI 10.1007/978-1-4614-5505-9_1, © The Author(s) 2013

These proteins, found on the surface of the most of the cells in the body, are known as human leukocyte antigens—HLA antigens—giving the body's immune system the ability to determine what belongs and what does not belong to the body. Whenever the immune system does not recognize the series of antigens on a cell surface, it creates antibodies and other substances to destroy the cells with non-recognizable antigens. These agents are also: infection-causing fungi, bacteria, viruses, tumor cells, and foreign agents, such as splinters. In this way, the immune system defends the body against potential invaders that can enter the body and cause harm. Regarding the bone marrow transplant to work, the recipient's immune system must not try to destroy the donated marrow. This comprises that the antigens—HLA system on the donated marrow cells have to be identical, or very similar to the antigens on the cells of the recipient.

In 1973 the first unrelated bone marrow transplantation on a 5-year-old patient suffering from severe combined immunodeficiency syndrome was performed by a team of physicians at Memorial Sloan-Kettering Cancer Center in New York City [3–6]. The donor was found in Denmark through the Blood Bank at Rigshospitalet in Copenhagen. The patient received multiple infusions of marrow and after the seventh transplant, engraftment was established and hematological function rapidly normalized. By this time, physicians had gained enough experience with bone marrow transplantation to determine that three types of antigens: HLA-A, HLA-B, and HLA-DR, were crucial in determining a match between donor and recipient. An individual inherits one of each of these three antigens from each of his/her parents. Therefore, a total of six antigens from the donor must match the six antigens of the recipient. Even with this careful matching, transplant may still fail because the recipient's immune system destroys the new marrow cells (called graft rejection), or because cells in the donor's marrow try to destroy the recipient's cells (called graft-versus-host disease, or GVHD).

In 1990, Dr. E. Donnall Thomas was awarded the Nobel Prize in Medicine for his pioneering work on transplantation [7–11]. During the early to mid-1970s, Thomas performed more than 100 transplants for patients with aplastic anemia and leukemia with HLA-A, HLA-B, and HLA-DR identical siblings. Because of Dr. Thomas's work and the dedication of many other physicians in the field of unrelated donor bone marrow transplantation, the procedure has evolved from an investigational therapy with uncertain benefit to a preferred first treatment for many patients [1–9].

Transplantation of Cord Blood and Adult Peripheral Blood Stem Cells

In 1988, successful transplantation occurred in a young boy with Fanconi anemia using umbilical cord blood collected at the birth of his sibling [10]. The patient remains alive and well to this date. In 1992, a patient was successfully transplanted with cord blood instead of bone marrow for the treatment of leukemia [10]. Over the past decade, the use of cord blood has expanded rapidly. Cord blood has been used to transplant in any disease state for which bone marrow can be used in spite of some disadvantage—rather small number of cells collected in each unit; delayed engraftment of neutrophils is common with cord blood. In addition to bone marrow and cord blood, peripheral blood stem cells (PBSCs) have gained popularity as a source of stem cells since their initial introduction in the 1980s.

Stem Cells Nowadays

The late 1990s brought a new apprehension regarding the biology of stem cells. In 1998, researchers at the University of Wisconsin led by James Thomson isolated and grew stem cells from human embryos, and researchers from Johns Hopkins University led by John Gearhart did the same for human germ cells [11]. In 1999 and 2000, researchers began to find that manipulation of adult mouse tissues could sometimes yield previously unsuspected cell types; for example, that some bone marrow cells could be turned into nerve or liver cells and that stem cells found in the brain appear to be able to form other kinds of cells.

The current knowledge related to the stem cells in 2005 comprises:

- Stem cells are the body's "master" cells.
- Stem cells can renew themselves indefinitely and differentiate into any of a number of types of specialized cells, such as muscles, nerves, organs, bone, and blood.
- Stem cells "plasticity"—ability to become other types of cells—makes them essential for repairing and renewing body tissues throughout our lives.

After we are born, our body retains stem cell reserves in various organs and, throughout our lives, we tap into those reserves to repair and replace injured or diseased tissues. Unfortunately, our stem cell reservoirs are finite and, as they become depleted, we succumb to diseases, disorders, and the ravages of aging. Thus, stem cell therapy offers the potential to replenish our exhausted reserves and fight a wide variety of diseases and disorders.

Summary

Stem cells are the key subset of cells in the body functioning as ancestor cells to produce a variety of types of functionally specialized mature cells (differentiation) in a given tissue, while at the same time maintaining the capacity to continuously divide and reproduce themselves (self-renewal). This self-renewal process is controlled by intrinsic genetic pathways that are subject to regulation by extrinsic signals from the microenvironment in which stem cells reside. Stem cells play essential roles ranging from embryonic development and organogenesis (fetal stem cells including embryonic stem cells) to tissue homeostasis and regeneration (adult stem cells). Stem cell development is a complex process: a precise balance is maintained among different cell events including self-renewal, differentiation, apoptosis (cell death), and migration. Loss of this balance tends to lead to uncontrolled cell growth/death, thereby developing into a variety of diseases including tissue defects or cancer. To investigate the molecular mechanisms that control stem cell properties we use the combined approaches described as follows:

- A global view of the changes in the gene expression patterns during hematopoietic stem cell development to reveal important pathways regulating hematopoietic stem cell self-renewal and lineage commitment. The Notch, Wnt, and BMP signal pathways have been well documented to be involved in cell fate determination during embryogenesis [11–14]. Expression of Notch, BMP4, and beta-Catenin (a transcription factor) in hematopoietic stem cells suggests that these pathways may play important roles in the regulation of hematopoietic stem cell proliferation and differentiation [3].
- To further characterize the functions of these pathways, we use genetic approaches such as transgenic or gene targeting animal models to examine their influence on stem cell development. Our goal is to understand how these signal pathways or mechanisms regulate normal development in the hematopoietic system. This information should reveal how they may malfunction or be altered in association with human diseases, such as leukemias, lymphomas, or autoimmune diseases.

References

1. Li L, Xie T (2005) Stem cell niche: structure and function. Annu Rev Cell Dev Biol 17:605–631
2. He XC, Zhang J, Li L (2005) Cellular and molecular regulation of hematopoietic and intestinal stem cell behavior. In: Stem cell biology: development and plasticity, vol 1049. Annals of the New York Academy of Sciences, New York, pp 28–38
3. Zhang J, Li L (2005) BMP signaling and stem cell regulation. Dev Biol 284:1–11
4. Li L (2005) Find the hematopoietic stem cell niche in placenta. Dev Cell 8:297–304
5. Tian Q, Feetham MC, Tao WA, He XC, Li L, Aebersold R, Hood L (2004) Proteomic analysis identifies that 14-3-3{zeta} interacts with {beta}-catenin and facilitates its activation by Akt. Proc Natl Acad Sci USA 101(43):15370–15375
6. He XC, Zhang J, Tong WG, Tawfik O, Ross J, Scoville DH, Tian Q, Zeng X, He X, Wiedemann LM, Mishina Y, Li L (2004) BMP signaling inhibits intestinal stem cell self-renewal through suppression of Wnt-beta-atenin signaling. Nat Genet 36:1117–1121
7. Zhang J, Niu C, Ye L, Huang H, He X, Tong WG, Ross J, Haug J, Johnson T, Feng JQ, Harri S, Wiedemann LM, Mishina Y, Li L (2003) Identification of the haematopoietic stem cell niche and control of the niche size. Nature 425:836–841
8. Akashi K, He X, Chen J, Iwasaki H, Niu C, Steenhard B, Zhang J, Haug J, Li L (2003) Transcriptional accessibility for genes of multiple tissues and hematopoietic lineages is hierarchically controlled during early hematopoiesis. Blood 101(2):383–389

9. Park I, He Y, Lin F, Laerum O, Tian Q, Bumgarner R, Klug C, Li K, Kuhr C, Doyle M, Xie X, Schummer M, Sun Y, Goldsmith A, Clarke M, Weissman I, Hood L, Li L (2002) Differential gene expression profiling of adult murine hematopoietic stem cells. Blood 99:488–498

10. Terskikh AV, Easterday MC, Li L, Hood L, Kornblum HI, Geschwind DH, Weissman IL (2001) From hematopoiesis to neuropoiesis: evidence of overlapping genetic programs. Proc Natl Acad Sci USA 98:7934–7939

11. Li L, Milner L, Deng Y, Iwata W, Banta A, Graf L, Marcovina S, Friedman C, Trask B, Hood L, Torok-Storb B (1998) The human homolog of rat Jagged1 expressed by marrow stroma inhibits differentiation of 32D cells through interaction with Notch1. Immunity 8:43–55

12. Li L, Krantz ID, Deng Y, Genin A, Banta A, Collins C, Qi M, Trask BJ, Kuo W, Cochran J, Costa T, Pierpont MEM, Rand EB, Piccoli D, Hood L, Spinner N (1997) Alagille syndrome is caused by mutations in hJagged1 (JAG1), which encodes a ligand for Notch1. Nat Genet 16:243–251

13. Lee TC, Li L, Philipson L, Ziff EB (1997) Myc represses transcription of the growth arrest gene gas-1. Proc Natl Acad Sci USA 94:12886–12891

14. Li L, Nerlov C, Prendergast G, MacGregor D, Ziff EB (1994) c-Myc represses transcription by a novel mechanism dependent on the initiator element and Myc box II. EMBO J 13: 4070–4079

Chapter 2
Stem Cell Concept: Entity or Function?

The ignorance, the root of every evil.

Plato

Stem cells derive from different sources:

- Embryonic tissues
- Fetal tissues
- Cord blood
- Adult tissues (different kind throughout the body)

They have the unique ability of self-renewal and plasticity, e.g., to multiply for a long time without a change and to produce cells that differentiate into specialized structures. New stem cells can be created to replenish stem cell population.

Genes that regulate these two distinct mechanisms are recently discovered:

1. The key gene that initiates the signaling process that instructs a stem cell to renew itself instead of differentiating into another type of cell is found in the model of drosophila (Erika Matunis and Natalia Tulina, from Department of Embryology at the Carnegie Institution in Washington-Science) [1].
2. The cascade of signals that start with the JAK-STAT signaling pathway where unpaired genes called Upd expressed in hub cells activating this cascade. When the pathway is activated, the STAT transcription factor—gene that controls the transcription of other genes—binds to target genes to change the cell's pattern of gene expression and help the cell to self-renew [2].

Depending on their origin they have varying capacity to multiply and differentiate from other cell types. It is not possible at present to predict which types of cells will be best suitable for various therapeutic situations. Embryonic stem cells are derived from pre-embryos at the blastocyst stage and may give rise to all bodily tissues and cells (have been shown capable of differentiating into all the different tissues and cell types of the body, and therefore have the highest, totipotent potential).

M. Pavlovic and B. Balint, *Stem Cells and Tissue Engineering*,
SpringerBriefs in Electrical and Computer Engineering,
DOI 10.1007/978-1-4614-5505-9_2, © The Author(s) 2013

How Does Stem Cell Differ from Other Cells?

Stem cells are unspecialized, capable of dividing and renewing themselves for long periods, and can give rise to specialized cells. Internal signals for both multiplication and differentiation are coming from genes, while external signals are limited to environment in their living compartments known as "niche." Stem cells have all of the genes that the other cells do, but they do not have the code to be any specific type of cell, neither many genes are "turned on." Their genes are activated for a primary purpose of mitosis without a specialized tissue function. In another words stem cells are cells that do not have a specific gene code "turned on" to be an exact, particular cell. Daughter cells derived from these stem cells are capable of differentiating into almost any kind of cell.

Types of Stem Cells: Entities and Functions

Within stem cell compartment there are different types of stem cells: embryonic stem cells are not only the most promising, but also the most controversial ones. When a sperm fertilizes an egg, it becomes what is known as a "zygote." Many scientists view the zygote as the ultimate stem cell because it can develop into any cell—not only of the embryo, but also of the surrounding tissues, such as placenta. The zygote has the highest degree of plasticity and it is referred to as a "totipotent" stem cell [1].

Thirty hours after fertilization, the zygote begins to divide, and by the fifth or sixth day, the cells form a kind of a bubble or "blastocyst." After the first week following fertilization, the cells begin to develop the coding sequence for specific functions, which makes isolating the stem cells during the blastocyst state imperative. When removed from the blastocyst, the cells can be cultured into "*embryonic stem cells,*" but these cells are not embryos. They have the capability of developing into all three types of tissue cells:

- Cells in the endoderm (which lines the digestive tract)
- Cells in the ectoderm (the outermost layer of the tissue cells)
- Cells in the mesoderm (fills the space between the endoderm and the ectoderm with cells such as muscle tissue)

These stem cells are somewhat less plastic and more specialized than totipotent zygote stem cells. Such stem cells that can become any of the more than 200 types of cells in the body are called "pluripotent" stem cells.

Between the seventh and ninth day, the blastocyst attaches to the uterus and begins to develop and grow. From this point until about 8 weeks it is generally referred to as an "embryo"; from 8 weeks on it is referred to as a "fetus."

Fetal stem cells. As the embryo grows it accumulates additional embryonic stem cells in yolk sack. From weeks 8 to 12 "fetal stem cells" are accumulated in the liver. Both embryonic and fetal stem cells generate the developing tissues and organs. At this stage such stem cells are designed as "multipotent," and they are more tissue-specific rather than generating all of the body's 200 different cell types [2]. Such stem cells are generally designated as "multipotent." However, some research suggests that at least some multipotent stem cells may be more plastic than first thought and may, under the right circumstances, become pluripotent.

Up until week 12, fetal stem cells (as well as the embryonic stem cells which preceded them) can be transplanted into an individual without being rejected. This is because they have little to none of immune-triggering proteins—HLA antigens on their surface.

After the 12th week, fetal stem cells acquire these proteins, and they remain present on stem cells from this point on, including on adult stem cells. Thus, while some advocate therapeutic use of stem cells derived from cord blood, adult bone marrow or the blood stream, these sources pose the problem of possible rejection reactions. Therefore, stem cells derived from these sources may have therapeutic potential only when given to the individual from whom they were derived ("autologous" transplantation) or from an immunologically matched donor ("allogenic" transplantation).

Adult stem cells are at a more advanced stage of development. For a long time adult stem cells were considered not capable of differentiating into the endoderm, ectoderm, or mesoderm, because they are already at a developed stage as one of the three types of tissues and cannot be rejuvenated back to an early developmental stage. They can be found in the blood, cornea, bone marrow, dental pulp of the tooth, brain, skeletal muscle, skin, liver, pancreas, and gastrointestinal tract [3]. These cells are capable of making identical copies of themselves, and usually divide to make "progenitor" or "precursor" cells capable to develop into specific cell lines. Adult stem cells have been identified in many organs and tissues, but in a very small number in each tissue. They are thought to reside in a specific area of each tissue (niche) where they may remain quiescent (non-dividing) for many years until they are activated by disease or tissue injury. The adult tissues reported to contain stem cells include brain, bone marrow, peripheral blood, blood vessels, skeletal muscle, olfactory mucosa, skin, and liver [4].

Adult stem cells typically generate the cell types of the tissue in which they reside. A blood-forming adult stem cell in the bone marrow, for example, normally gives rise to the many types of blood cells such as erythrocytes, granulocytes, lymphocytes, and platelets. Until recently, it had been thought that a blood-forming cell in the bone marrow—which is called a hematopoetic stem cell—could not give rise to the cells of a very different tissue, such as nerve cells in the brain. However, a number of experiments over the last several years have raised the possibility that stem cells from one tissue may be able to give rise to cell types of a completely different tissue—a phenomenon known as plasticity [3]. Examples of such plasticity include blood cells becoming neurons, liver cells that can be made to produce insulin and hematopoietic stem cells that can develop into heart muscle. Furthermore, the concept of plasticity has been revised by the Ratajczak's group which has developed recently and proved the concept of very small embryonic like cell (VSEL), shown to be stem cells in bone marrow and other organs in non-hematopoietic compartment, committed to differentiate into some other tissues [4]. Therefore, exploring the possibility of using adult stem cells for cell-based therapies has become a very active area of investigation. As an adult stem cell is an undifferentiated cell found among differentiated cells in a tissue or organ, it can renew itself and can differentiate to yield the major specialized cell types of the tissue or organ. The primary role of adult stem cells in a living organism is to maintain and repair the tissue in which they reside. Unlike embryonic stem cells which are defined by their origin (the inner mass of the blastocyst), the origin of adult stem cells in mature tissues is unknown. Research on adult stem cells has recently generated a great deal of excitement. Could adult stem cells be used for transplants? In fact, adult blood-forming stem cells from bone marrow have been used in transplants for 30 years. Certain kinds of adult stem cells seem to have the ability to differentiate into a number of different cell types, given the right conditions [5]. If this differentiation of adult stem cells can be controlled in the laboratory, these cells may become the basis of therapies for many serious common diseases, and they are already becoming, in some of them [3]. For example, osteogenesis imperfecta treated with mesenchymal stem cells has a tremendous outcome within children.

References

1. Tulina N, Matunis E (2001) Control of stem cell self-renewal in Drosophila spermatogenesis by JAK-STAT signaling. Science 294(5551):2546–2549
2. Tran J, Brenner JT, DiNardo S (2000) Somatic control over the germline stem cell lineage during *Drosophila* spermatogenesis. Nature 407:754–757. doi:10.1038/35037613
3. Balint B, Stamatovic D, Todorovic M, Jevtic M, Ostojic G, Pavlovic M et al (2007) Stem cells in the arrangement of bone marrow repopulation and regenerative medicine. Vojnosanit Pregl 64(7):481–484
4. Ratajczak MZ, Zuba-Surma EK, Ratajczak J, Wysoczynski M, Kucia M (2008) Very small embryonic like (VSEL) stem cells—characterization, developmental origin and biological significance. Exp Hematol 36(6):742–751
5. Atala A, Lanza R (eds) (2001) Methods of tissue engineering, San Diego, CA, Academic Press

Chapter 3
Embryonic Stem Cells: Problems and Possible Solutions

Glory is fleeting, but obscurity is forever.

- Napoleon Bonaparte

Animal models have demonstrated that transplanted embryonic cells are exposed to the immune reactions similar to those acting on organ transplants, hence immuno-suppression of the recipient is generally required. It is, however, possible to obtain embryonic stem cells that are genetically identical to the patient's own cells. The nucleus from the patient's somatic cell is transferred into an egg after removal of the egg's own genetic material (a technique known as nuclear transfer or therapeutic cloning). Under specific condition the egg will use genetic information from the patient's somatic cell in organizing the formation of a blastocyst which in turn generates embryonic stem cells. These cells have a genetic composition identical to that of patient, are suitable for stem cell therapy, will generate patient's own proteins, and escape the danger for "self-attack" and immune rejection [1].

What is other problem about embryonic stem cells? Stem cells are the primordial goop of the human body cells that have not yet been differentiated into bone, blood, or brain cells. For medical researchers, stem cells represent a mother lode of possible new treatments for diabetes, heart disease, Parkinson's, and more. Capable of differentiating into the full spectrum of other cell types—from a new liver cell to a new neuron—they could be ideal for repairing or replacing diseased organs. They are the basic and critical entity for establishment and development of tissue engineering.

The furor over stem-cell research is over their source: should researchers use aborted or discarded human embryos? Or should they be restricted to adult stem cells, found in fat, bone, the brain, and other sources? Because of ethical problems only embryos which cannot be used in fertility treatment, and have been donated for research, can be used, so far. The ethical questions are certainly riveting, but they may be swiftly trumped by the market, specifically the venture investment market, which is voting with its dollars for adult stem-cell research. Why? If you look at

M. Pavlovic and B. Balint, *Stem Cells and Tissue Engineering*,
SpringerBriefs in Electrical and Computer Engineering,
DOI 10.1007/978-1-4614-5505-9_3, © The Author(s) 2013

some of the medical and scientific indications, adult stem cells are much closer to therapeutic applications; embryonic cells still have a variety of obstacles that need to be overcome. Moreover, embryo cells can be a bit too flexible, differentiating into all kinds of tissue, both desirable and not. When injected under the skin of certain mice, for example, they grow into teratomas, tumors consisting of numerous tissue types, from gut to skin to teeth. Injected adult stem cells are better behaved, growing into other tissues only after the application of appropriate growth factors or other external cues. This need for external cues is ideal since doctors want to be able to tightly control results [2, 3].

Overview

James Thomson (1988) with colleagues reported methods for deriving and maintaining human embryonic stem (ES) cells from the inner cell mass of human blastocysts that were produced through in vitro fertilization (IVF) and donated for research purposes [4]. John Gearhar's group reported at the same time the derivation of cells that were identified as embryonic germ (EG) cells. The cells were cultured from primordial germ cells obtained from the gonadal ridge and mesenchyma of 5- to 9-week fetal tissue that resulted from elective abortions [5].

The methods were developed by two teams for culturing human ES and EG cells by drawing on a host of animal studies based in the past: derivations of pluripotent mouse ES cells from blastocysts (Andrews PW, personal communication) [6], reports of the derivation of EG cells [7, 8], experiments with stem cells derived from mouse teratocarcinomas [9–11], the derivation and culture of ES cells from the blastocysts of rhesus monkeys [4] and marmosets [12], and methods used by IVF clinics to prepare human embryos for transplanting into the uterus to produce a live birth [13, 14].

Since the first isolation of embryonic stem cells from human blastocysts in 1994 (Andrews PW, personal communication; Bongso A, personal communication) [7–14], techniques for deriving and culturing human ES cells have been refined [15, 16]. The ability to isolate human ES cells from blastocysts and grow them in culture seems to depend in large part on the integrity and condition of the blastocyst from which the cells are derived. In general, blastocysts with a large and distinct inner cell mass tend to yield ES cultures most efficiently [13].

Blastocyst In Vitro

After a human oocyte is fertilized in vitro by a sperm cell, the cascade of events occur in fertilized egg (zygote) according to a fairly predictable timeline (Bongso A, personal communication) [15–19]. A normal day-5 human embryo in vitro consists of 200–250 cells, when blastocysts are used to derive ES cell cultures. Most of the cells

Fig. 3.1 Human blastocyst showing inner cell mass and trophectoderm (photo credit: Mr. J. Conaghan)

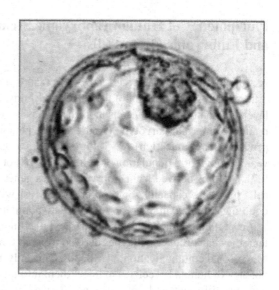

comprise the trophectoderm. For deriving ES cell cultures, the trophectoderm is removed, either by microsurgery or immunosurgery (in which antibodies against the trophectoderm help break it down, thus freeing the inner cell mass, composed of only 30–34 cells (Bongso A, personal communication) [9–20]. The in vitro conditions for growing a human embryo to the blastocyst stage vary among IVF clinics and are reviewed elsewhere (Gearhart J, personal communication) [14, 21–26]. However, once the inner cell mass is obtained from either mouse or human blastocysts, the techniques for growing ES cells are similar (Fig. 3.1).

Derivation of Human Embryonic Germ Cells

Human embryonic germ (EG) cells share many of the characteristics of human ES cells, but differ in significant ways. Human EG cells derive from the primordial germ cells, in a specific part of the embryo/fetus (gonadal ridge), which normally develops into mature gametes (eggs and sperm). Gearhart and his collaborators devised methods for growing pluripotent cells derived from human EG cells [17]. The process requires the generation of embryoid bodies from EG cells, which consists of an unpredictable mix of partially differentiated cell types (Gearhart J, personal communication) [17–27]. The embryoid body-derived cells resulting from this process have high proliferative capacity and gene expression patterns that are representative of multiple cell lineages, suggesting that the embryoid body derived cells are progenitor or precursor cells for a variety of differentiated cell types [27].

Pluripotency of Human Embryonic Stem Cells and Embryonic Germ Cells

A truly pluripotent stem cell is a cell that is capable of self-renewal and of differentiating into most all of the cells of the body, including cells of all three germ layers. Human ES and EG cells in vitro are capable of long-term self-renewal, while retaining a normal karyotype (Andrews PW, personal communication; Bongso A, personal communication) [1, 5–15, 28]. Human ES cells can proliferate for two years through 300 population doublings [26–28] or even 450 population doublings [27]. Cultures derived from embryoid bodies generated by human embryonic germ cells have less capacity for proliferation. Most will proliferate for 40 population doublings; the maximum reported is 70–80 population doublings [29].

Several laboratories have demonstrated that human ES cells in vitro are pluripotent; they can produce cell types derived from all three embryonic germ layers (Pera M, personal communication) [1, 18, 30]. Currently, the only test of the in vivo pluripotency of human ES cells is to inject them into immune-deficient mice where they generate differentiated cells that are derived from all three germ layers. These include gut epithelium (which, in the embryo, is derived from endoderm); smooth and striated muscle (derived from mesoderm); and neural epithelium, and stratified squamous epithelium (derived from ectoderm; Pera M, personal communication) [4, 18, 29–37]. Currently, a major goal for embryonic stem cell research is to control the differentiation of human ES and EG cell lines into specific kinds of cells—an objective that must be met if the cells are to be used as the basis for therapeutic transplantation, testing drugs, or screening potential toxins. The techniques now being tested to direct human ES cell differentiation are borrowed directly from techniques used to direct the differentiation of mouse ES cells in vitro.

What Are Potential Uses of Human Embryonic Stem Cells in Tissue Engineering?

Many uses have been proposed for human embryonic stem cells. The most-often discussed is their potential use in transplant therapy or tissue engineering—i.e., to replace or restore tissue that has been damaged by disease or injury.

Using Human Embryonic Stem Cells for Therapeutic Transplants

Diseases that might be treated by transplanting human ES-derived cells include Parkinson's disease, diabetes, traumatic spinal cord injury, Purkinje cell degeneration, Duchenne's muscular dystrophy, heart failure, and osteogenesis imperfecta.

The research is occurring in several laboratories, but is limited because so few laboratories have access to human ES cells. Thus, at this stage, many therapies based on the use of human ES cells are still hypothetical and highly experimental [20, 26, 28].

One of the *current advantages* of using ES cells as compared to adult stem cells is that ES cells have an unlimited ability to proliferate in vitro and are more likely to be able to generate a broad range of cell types through directed differentiation. Ultimately, it will also be necessary to both identify the optimal stage(s) of differentiation for transplant and demonstrate that the transplanted ES-derived cells can survive, integrate, and function in the recipient.

The potential *disadvantages* of the use of human ES cells for transplant therapy include the propensity of undifferentiated ES cells to induce the formation of tumors (teratomas), which are typically benign. Because it is the undifferentiated cells— rather than their differentiated progeny—that have been shown to induce teratomas, tumor formation might be avoided by devising methods for removing any undifferentiated ES cells prior to transplant. Also, it should be possible to devise a fail-safe mechanism—i.e., to insert into transplanted ES-derived cells suicide genes that can trigger the death of the cells should they become tumorigenic. The potential immunological rejection of human ES-derived cells might be avoided by genetically engineering the ES cells to express the MHC antigens of the transplant recipient, or by using nuclear transfer technology to generate ES cells that are genetically identical to the person who receives the transplant, as already mentioned [26].

What Are Other Potential Uses of Human Embryonic Stem Cells?

Many potential uses of human ES cells have been proposed that do not involve transplantation. For example, human ES cells could be used to study early events in human development and so make it possible to identify the genetic, molecular, and cellular events that lead to birth defect problems and identify methods for their prevention [20, 35, 38]. They could also be used to explore the effects of chromosomal abnormalities in early development, including ability to monitor the development of early childhood tumors, many of which are embryonic in origin [39].

Testing candidate therapeutic drugs and even toxins by using ES cells involves human cell lines. Although animal model testing is a mainstay of pharmaceutical research, it cannot always predict the effects that a candidate drug may have on human cells. For this reason, cultures of human cells are often employed in preclinical tests. These human cell lines have usually been maintained in vitro for long periods and as such often have different characteristics than do in vivo cells. These differences can make it difficult to predict the action of a drug in vivo based on the response of human cell lines in vitro. Therefore, if human ES cells can be directed to differentiate into specific cell types that are important for drug screening, the ES-derived cells may be more likely to mimic the in vivo response of the cells/tissues to the drug(s) being tested and so offer safer, and potentially cheaper, models for drug screening. The reasons for using human ES cells to screen potential toxins closely resemble those for using human ES-derived cells to test drugs. Toxins often have different effects on different animal species, which makes it critical to have the best possible in vitro models for evaluating their effects on human cells. Finally, human ES cells could be used to develop new methods for genetic engineering. Currently, the genetic complement of mouse ES cells in vitro can be modified easily by techniques such as homologous recombination. This is a method for replacing or adding genes, which requires a DNA molecule artificially introduced into the genome and then expressed. Using this method, genes to direct differentiation to a specific cell type or genes that express a desired protein product might be introduced into the ES cell line. Ultimately, if such techniques could be developed using human ES cells, it may be possible to devise better methods for gene therapy [38].

What Are Critical Questions Regarding Human Embryonic Stem Cells?

There is the difference between ES cells and human EG cells which are apparently not equivalent in their potential to proliferate or differentiate. Both kinds of cells spontaneously generate neural precursor-type cells (widely regarded as a default pathway for differentiation), both generating cells that resemble cardiac myocytes [17, 35]. However, ES cells, derived from the inner cell mass of the preimplantation blastocyst, approximately 5 days post-fertilization, should be distinguished from human EG cells, derived from fetal primordial germ cells, 5–10 weeks post-fertilization. ES cells can proliferate for up to 300 population doublings, while cells derived from embryoid bodies that are generated from embryonic germ cells (fetal tissue) double a maximum of 70–80 times in vitro. ES cells appear to have a broader ability to differentiate. Both human ES and EG cells in vitro will spontaneously generate embryoid bodies that consist of cell types from all three primary germ layers (Pera M, personal communication) [1, 18, 29]. Many key questions regarding ES remain unanswered or only partly answered. For example, in order to refine and improve ES cell culture systems, we have to identify the mechanisms that allow human ES cells in vitro to proliferate without differentiating [26]. Once the mechanisms that regulate human ES proliferation are completely known, it will be highly likely to apply this knowledge to the long-standing challenge of improving the in vitro self-renewal capabilities of adult stem cells.

It is of critical significance to determine whether the genetic imprinting status of human ES cells plays any significant role in maintaining the cells, directing their differentiation, or determining their suitability for transplant. One of the effects of growing mouse blastocysts in culture is a change in the methylation of specific genes that control embryonic growth and development [21]. Do similar changes in gene imprinting patterns occur in human ES cells (or blastocysts)? If so, what is their effect on in vitro development and on any differentiated cell types that may be derived from cultured ES cells?

The fundamental importance is in determining whether cultures of human ES cells that appear to be homogeneous and undifferentiated are, in fact, homogeneous and undifferentiated. Is it possible that human ES cells in vitro cycle in and out of partially differentiated states? And if that occurs, how will it affect attempts to direct their differentiation or maintain the cells in a proliferating state? [25]. There is a need to identify which signal transduction pathways must be activated to induce human ES cell differentiation along a particular pathway. This includes understanding ligand–receptor interaction and the intracellular components of the signaling system, as well as identifying the genes that are activated or inactivated during differentiation of specific cell types [26]. Identifying intermediate stages of human ES cell differentiation will also be important. As human ES cells differentiate in vitro, do they form distinct precursor or progenitor cells that can be identified and isolated? Would such precursor or progenitor cells be useful for therapeutic transplantation? [17]. Finally, we need to determine what differentiation stages of human ES-derived cells are optimal for particular practical applications. For example, what differentiation stages of ES-derived cells would be best for screening drugs or toxins, or for delivering potentially therapeutic drugs?

References

1. Amit M, Carpenter MK, Inokuma MS, Chiu CP, Harris CP, Waknitz MA, Itskovitz-Eldor J, Thomson JA (2000) Clonally derived human embryonic stem cell lines maintain pluripotency and proliferative potential for prolonged periods of culture. Dev Biol 227:271–278
2. Andrews PW, Damjanov I, Simon D, Banting GS, Carlin C, Dracopoli NC, Fogh J (1984) Pluripotent embryonal carcinoma clones derived from the human teratocarcinoma cell line Tera-2. Differentiation in vivo and in vitro. Lab Invest 50:147–162
3. Andrews PW (1988) Human teratocarcinomas. Biochim Biophys Acta 948:17–36
4. Thomson JA, Kalishman J, Golos TG, Durning M, Harris CP, Hearn JP (1996) Pluripotent cell lines derived from common marmoset (Callithrix jacchus) blastocysts. Biol Reprod 55: 254–259
5. Andrews PW (1998) Teratocarcinomas and human embryology: pluripotent human EC cell lines. Review article. APMIS 106:158–167
6. Shamblott MJ, Axelman J, Wang S, Bugg EM, Littlefield JW, Donovan PJ, Blumenthal PD, Huggins GR, Gearhart JD (1998) Derivation of pluripotent stem cells from cultured human primordial germ cells. Proc Natl Acad Sci USA 95:13726–13731
7. Bongso A, Fong CY, Ng SC, Ratnam SS (1994) Blastocyst transfer in human in vitro: fertilization; the use of embryo co-culture. Cell Biol Int 18:1181–1189
8. Bongso A, Fong CY, Ng SC, Ratnam S (1994) Isolation and culture of inner cell mass cells from human blastocysts. Hum Reprod 9:2110–2117
9. Bongso A, Fong CY, Ng SC, Ratnam SS (1995) Co-culture techniques for blastocyst transfer and embryonic stem cell production. Asst Reprod Rev 5:106–114
10. Bongso A (1996) Behaviour of human embryos in vitro in the first 14 days: blastocyst transfer and embryonic stem cell production. Clin Sci 91:248–249
11. Kleinsmith LJ, Pierce GB Jr (1964) Multipotentiality of single embryonal carcinoma cells. Cancer Res 24:1544–1551
12. Thompson S, Stern PL, Webb M, Walsh FS, Engstrom W, Evans EP, Shi WK, Hopkins B, Graham CF (1984) Cloned human teratoma cells differentiate into neuron-like cells and other cell types in retinoic acid. J Cell Sci 72:37–64

13. Bongso A (1999) Handbook on blastocyst culture. Sydney Press Indusprint, Singapore
14. Trounson AO, Gardner DK, Baker G, Barnes FL, Bongso A, Bourne H, Calderon I, Cohen J, Dawson K, Eldar-Geve T, Gardner DK, Graves G, Healy D, Lane M, Leese HJ, Leeton J, Levron J, Liu DY, MacLachlan V, Munné S, Oranratnachai A, Rogers P, Rombauts L, Sakkas D, Sathananthan AH, Schimmel T, Shaw J, Trounson AO, Van Steirteghem A, Willadsen S, Wood C (2000) Handbook of in vitro fertilization. CRC, Boca Raton
15. Reubinoff BE, Pera MF, Fong CY, Trounson A, Bongso A (2000) Embryonic stem cell lines from human blastocysts: somatic differentiation in vitro. Nat Biotechnol 18:399–404
16. Trounson AO, Anderiesz C, Jones G (2001) Maturation of human oocytes in vitro and their developmental competence. Reproduction 121:51–75
17. Bradley A, Evans M, Kaufman MH, Robertson E (1984) Formation of germ-line chimaeras from embryo-derived teratocarcinoma cell lines. Nature 309:255–256
18. De Vos A, Van Steirteghem A (2000) Zona hardening, zona drilling and assisted hatching: new achievements in assisted reproduction. Cells Tissues Organs 166:220–227
19. Evans MJ, Kaufman MH (1981) Establishment in culture of pluripotential cells from mouse embryos. Nature 292:154–156
20. Fong CY, Bongso A, Ng SC, Kumar J, Trounson A, Ratnam S (1998) Blastocyst transfer after enzymatic treatment of the zona pellucida: improving in-vitro fertilization and understanding implantation. Hum Reprod 13:2926–2932
21. Friedrich TD, Regenass U, Stevens LC (1983) Mouse genital ridges in organ culture: the effects of temperature on maturation and experimental induction of teratocarcinogenesis. Differentiation 24:60–64
22. Gardner DK, Schoolcraft WB (1999) Culture and transfer of human blastocysts. Curr Opin Obstet Gynecol 11:307–311
23. Itskovitz-Eldor J, Schuldiner M, Karsenti D, Eden A, Yanuka O, Amit M, Soreq H, Benvenisty N (2000) Differentiation of human embryonic stem cells into embryoid bodies comprising the three embryonic germ layers. Mol Med 6:88–95
24. Jones GM, Trounson AO, Lolatgis N, Wood C (1998) Factors affecting the success of human blastocyst development and pregnancy following in vitro fertilization and embryo transfer. Fertil Steril 70:1022–1029
25. Jones JM, Thomson JA (2000) Human embryonic stem cell technology. Semin Reprod Med 18:219–223
26. Khosla S, Dean W, Brown D, Reik W, Feil R (2001) Culture of preimplantation mouse embryos affects fetal development and the expression of imprinted genes. Biol Reprod 64:918–926
27. Bongso A, Fong CY, Mathew J, Ng LC, Kumar J, Ng SC (1999) The benefits to human IVF by transferring embryos after the in vitro embryonic block: alternatives to day 2 transfers. Asst Reprod Rev 9:70–78
28. Martin GR (1980) Teratocarcinomas and mammalian embryogenesis. Science 209:768–776
29. Resnick JL, Bixler LS, Cheng L, Donovan PJ (1992) Long-term proliferation of mouse primordial germ cells in culture. Nature 359:550–551
30. Pera MF, Reubinoff B, Trounson A (2000) Human embryonic stem cells. J Cell Sci 113 (Pt 1):5–10
31. Pera MF, Cooper S, Mills J, Parrington JM (1989) Isolation and characterization of a multipotentclone of human embryonal carcinoma cells. Differentiation 42:10–23
32. Rathjen PD, Lake J, Whyatt LM, Bettess MD, Rathjen J (1998) Properties and uses of embryonic stem cells: prospects for application to human biology and gene therapy. Reprod Fertil Dev 10:31–47
33. Reubinoff BE, Pera M, Fong CY, Trounson A, Bongso A (2000) Research errata. Nat Biotechnol 18:559
34. Sathananthan AH (1997) Ultrastructure of the human egg. Hum Cell 10:21–38
35. Schuldiner M, Yanuka O, Itskovitz-Eldor J, Melton D, Benvenisty N (2000) Effects of eight growth factors on the differentiation of cells derived from human embryonic stem cells. Proc Natl Acad Sci USA 97:11307–11312

36. Shamblott MJ, Axelman J, Littlefield JW, Blumenthal PD, Huggins GR, Cui Y, Cheng L, Gearhart JD (2001) Human embryonic germ cell derivatives express a broad range of developmentally distinct markers and proliferate extensively in vitro. Proc Natl Acad Sci USA 98:113–118

37. Smith AG (2001) Origins and properties of mouse embryonic stem cells. Annu Rev Cell Dev Biol

38. Odorico JS, Kaufman DS, Thomson JA (2001) Multilineage differentiation from human embryonic stem cell lines. Stem Cells 19:193–204

39. Martin GR (1981) Isolation of a pluripotent cell line from early mouse embryos cultured in medium conditioned by teratocarcinoma stem cells. Proc Natl Acad Sci USA 78:7634–7638

Chapter 4
Adult Stem Cells (the Concept of VSEL-Cell)

You cannot teach a man anything; you can only find it within yourself.

– Galileo Galilei

Adult stem cells are stem cells that can be derived from different parts of the body and, depending on where they are from, have different properties. They exist in several different tissues including bone marrow, blood, liver, nasal mucosa, skin, and the brain. Some studies have suggested that adult stem cells are very versatile and can develop into many different cell types. Adult stem cells have already been used for more than 20 years as bone-marrow transplants to reconstitute the immune systems of patients with cancer and to treat blood cancers such as leukemia [1, 2]. Using the body's own stem cells means, the immune system's rejection reflex will not be aroused.

However, other studies have concluded that adult stem cells have limited differentiation potential. Although a wealth of information on adult stem cells has already accumulated, scientists still do not understand their specific properties very well [2, 3]. Research continues with the hope of 1 day being able to use these cells to restore or replace damaged tissues or organs. The VSEL concept, based on Dr Ratajczak's group studies, has radically changed the knowledge on adult stem cells and opened the door to new approach in stem cell treatment of different diseases [3, 4]. These scientists have found that beside hematological, there is a non-hematological pool of adult stem cells in the bone marrow and possible other organs, which are committed to differentiate in a certain type of cell [3, 4]. They do remind of embryonic stem cells since they have embryonic body in the cytoplasm, but they do have most stigmata of adult stem cells and can be mobilized by G-CSF into peripheral blood [2]. The future work is to show the meritory value of this concept and its applicability in clinical arena.

M. Pavlovic and B. Balint, *Stem Cells and Tissue Engineering*,
SpringerBriefs in Electrical and Computer Engineering,
DOI 10.1007/978-1-4614-5505-9_4, © The Author(s) 2013

References

1. Balint B, Stamatovic D, Todorovic M, Jevtic M, Ostojic G, Pavlovic M et al (2007) Stem cells
   in the arrangement of bone marrow repopulation and regenerative medicine. Vojnosanit Pregl
   64(7):481–484
2. Wojakowski W, Kucia M, Liu R, Zuba-Surma E, Jadczyk T, Bachowski R, Nabiałek E,
   Kaźmierski M, Ratajczak MZ, Tendera M (2011) Circulating very small embryonic-like stem
   cells in cardiovascular disease. J Cardiovascular Transl Res 4(2):138–144
3. VSEL technology and the future of regenerative medicine: an interview with Dr. Mariusz
   Dr. Mariusz Ratajczak (2012) H+ Community Magazine
4. Ratajczak MZ, Zuba-Surma EK, Ratajczak J, Wysoczynski M, Kucia M (2008) Very small
   embryonic like (VSEL) stem cells—characterization, developmental origin and biological
   significance. Exp Hematol 36(6):742–751. doi:10.1016/j.exphem.2008.03.010

Chapter 5
Cord Blood Stem Cells

Not everything that can be counted counts,
and not everything that counts can be counted.

– Albert Einstein

Like bone marrow, umbilical cord blood is another rich source of hematopoietic stem cells, being less mature than those found in the bone marrow of adults or children.

Advantages: The advantages of using cord blood as a source of stem cells are: its noninvasive procurement and its vast abundance. Therefore, cord blood is world-wide collected and either banked in public banks for general use, or stored by private companies for private use. Cord blood has recently emerged as an alternative source of hematopoietic stem cells for treatment of leukemia and other blood disorders. The notable advantage of this application is that despite its high content of immune cells, it does not produce strong graft-versus-host disease. Therefore, cord blood grafts do not need to be as rigorously matched to a recipient as bone marrow grafts [1]. This expands the available donor pool for hematopoietic stem cell transplants considerably. Cord blood cells are more abundant than bone marrow cells and are easier to collect than bone marrow. Collecting and storing of one's own baby's cord blood stem cells does not present an ethical dilemma because they come from the umbilical cord after delivery of the baby. Finally, cord blood collection is painless, noninvasive process.

Disadvantages: However, a disadvantage, and an argument against generalized use, is the limited number of stem cells in any given cord [2]. A typical cord blood harvest only contains enough stem cells to transplant a large child or small adult (weighing approximately 100 pounds). This is critical for the risk of graft failure once transplanted into an adult. Because the stem cells in the cord blood are more primitive than those in the bone marrow, the engraftment process takes longer with cord blood, leaving the patient vulnerable to a fatal infection for a longer period of

M. Pavlovic and B. Balint, *Stem Cells and Tissue Engineering*,
SpringerBriefs in Electrical and Computer Engineering,
DOI 10.1007/978-1-4614-5505-9_5, © The Author(s) 2013

time. Current research is exploring the methods and safety of transplanting adults with cord blood [3, 4].

The use of umbilical cord blood stem cells for other uses, such as organ and tissue repair, is under investigation. Current research involves the possible use of cord blood stem cells for the treatment of other non-blood diseases such as heart disease.

References

1. Vendrame M et al (2005) Anti-inflammatory effects of human cord blood cells in a rat model of stroke. Stem Cells Dev 14(5):595–604
2. Revoltella RP et al (2008) Cochlear repair by transplantation of human cord blood CD133+ cells to nod-scid mice made deaf with kanamycin and noise. Cell Transplant 17(6):665–678
3. Harris DT et al (2007) The potential of cord blood stem cells for use in regenerative medicine. Expert Opin Biol Ther 7(9):1311–1322
4. Haller MJ et al (2008) Autologous umbilical cord blood infusion for type 1 diabetes. Exp Hematol 36(6):710–715

Chapter 6
Hematopoietic Stem Cells

*I do not feel obliged to believe that the same
God who has endowed us with sense, reason,
and intellect has intended us to forgot their use.*

– Galileo Galilei

Hematopoietic stem cells are adult stem cells found mainly in the bone marrow and they provide the blood cells required for daily blood turnover and for fighting infections. Compared to adult stem cells from other tissues, hematopoietic stem cells are easy to obtain, as they can be either aspirated directly out of the bone marrow or stimulated to move into the peripheral blood stream, where they can be easily collected as shown by Balint et al. [1]. As previously mentioned, hematopoietic stem cells have been studied by scientists for many years, and they were the first stem cells to be used successfully in therapies, e.g., treatment of blood cancers (leukemia) and other blood disorders [2]. More recently, their use in treatment of breast cancer and coronary artery diseases has also been explored [3]. The potential for hematopoietic stem cells to produce cell types other than blood cells has become the subject of intense scientific controversy, and it is still not clear whether they could be used on a clinical scale to restore tissues and organs other than blood and the immune system. Their features will be described within few chapters of this book.

M. Pavlovic and B. Balint, *Stem Cells and Tissue Engineering*,
SpringerBriefs in Electrical and Computer Engineering,
DOI 10.1007/978-1-4614-5505-9_6, © The Author(s) 2013

References

1. Balint B, Stamatovic D, Andric Z (2003) Stem and progenitor cell transplantation. In: Balint B (ed) Transfusion Medicine. CTCI, Belgrade, pp 525–547
2. Ivanovic Z, Kovacevic-Filipovic M, Jeanne M, Ardilouze L, Bertot A, ne Szyporta M, Hermitte F, Lafarge X, Duchez P, Vlask M, Milpied M, Pavlovic M, Praloran V, Boiron J-M (2010) CD34+ cells obtained from âgood mobilizersâ are more activated and exhibit lower ex vivo expansion efficiency than their counterparts from âpoor mobilizersâ. Transfusion 50(1): 120–127
3. Balint B, Stamatovic D, Todorovic M, Elez M, Vojvodic D, Pavlovic M, Cucuz–Jokic M (2011) Autologous transplant in aplastic anemia: quantity of CD34$^+$/CD90$^+$ subset as the predictor of clinical outcome. Transfus Apher Sci 45(2):137–141

Chapter 7
Ethical Aspects of Stem Cell Research

The important thing in science is not so much to obtain
new facts as to discover new ways of thinking about them.

– Sir William Lawrence Bragg

Stem cell biology is an extremely active field in biology, not only from a scientific but also from the political, social, and ethical perspective. It is well known that ethical aspect involves for many strong argument that using embryonic stem cells is equal to homicide. The other argument is: How one can give embryonic stem cells to somebody, when the immune system will not tolerate "nonself"? This obstacle was overcome by the concept of therapeutic cloning, in which the nucleus of stem cell (through microsurgery) is replaced by patient's nucleus (nuclear transfer). In that way, embryonic body (which is in cytoplasm of the zygot and will give stem cells) will be "supervised" by patient's genetic codes and orchestrate patient's coding of specific proteins. They will not be foreign to immune system and will not affect the engraftment.

Despite scientific solution in order to get the most potent multipotent stem cells, which can diversify indefinitely, the first argument is still socially and politically significantly powerful. Meanwhile, adult stem cells have been found in different sources of the body and are more and more successfully used in the treatment of a number of diseases [1]. Embryonic stem cells are recently used in the case of two different types of macular degeneration in Harvard Medical School, and have shown encouraging results [2]. The time is necessary to pass and evaluate both practical and ethical issues.

M. Pavlovic and B. Balint, *Stem Cells and Tissue Engineering*,
SpringerBriefs in Electrical and Computer Engineering,
DOI 10.1007/978-1-4614-5505-9_7, © The Author(s) 2013

References

1. Balint B, Stamatovic D, Todorovic M, Elez M, Vojvodic D, Pavlovic M, Cucuz–Jokic M (2011) Autologous transplant in aplastic anemia: quantity of CD34+/CD90+ subset as the predictor of clinical outcome. Transfus Apher Sci 45(2):137–141
2. Schwartz SD, Hubschman J-P, Heilwell G, Franco-Cardenas V, Pan CK, Ostrick RM, Mickunas E, Gay R, Klimanskaya I, Lanza R (2102) Embryonic stem cell trial for macular degeneration: a preliminary report. Lancet 379(9817):713–720

Chapter 8
Stem Cell Renewal and Differentiation

This book fills a much-needed gap.

– Moses Hadas

Among the most investigated are the mechanisms that regulate stem cell function in the nervous and hematopoietic systems. Therefore, hematopoietic stem cells, which give rise to all blood and immune system cells, and neural crest stem cells, which give rise to the peripheral nervous system, are among the best-characterized stem cells. We are just beginning to understand how their functions are regulated. The conserved mechanisms have long been hypothesized as the mode of stem cell regulation. However, testing this requires interdisciplinary approaches. The ultimate goal is to integrate what we know about stem cells in different tissues to understand the extent to which they employ similar or different mechanisms to regulate critical functions. We have focused so far on the mechanisms that regulate stem cell self-renewal, aging, and their role in organogenesis.

Stem Cell Self-Renewal: Regulation

Persistence of stem cells which are maintained in numerous adult tissues by self-renewal (stem cells dividing to make more stem cells) raises the question of whether this process is regulated by mechanisms that are conserved between tissues.

It was found that the polycomb family transcriptional repressor Bmi-1 is required for the self-renewal but not for the differentiation of stem cells in the hematopoietic system and peripheral and central nervous systems [1]. In each case, stem cells are formed in normal numbers during fetal development but exhibit impaired self-renewal and become depleted postnatally. Bmi-1 promotes stem cell self-renewal partly by repressing $p16^{Ink4a}$, a cyclin-dependent kinase inhibitor, and $p19^{Arf}$, a p53 agonist. Both of these checkpoint proteins negatively regulate cell proliferation, and

M. Pavlovic and B. Balint, *Stem Cells and Tissue Engineering*,
SpringerBriefs in Electrical and Computer Engineering,
DOI 10.1007/978-1-4614-5505-9_8, © The Author(s) 2013

their increased expression has been associated with cellular senescence. This demonstrates that stem cells require mechanisms to prevent premature senescence in order to self-renew throughout adult life. In contrast, restricted neural progenitors from the enteric nervous system and forebrain proliferate normally in the absence of Bmi-1 [2]. Thus Bmi-1 dependence is conserved between stem cells and distinguishes the cell cycle regulation of stem cells from the cell cycle regulation of at least some types of restricted progenitors. Using similar approaches, studying of additional pathways could be done that as hypothesized, that will also regulate stem cell self-renewal, and that will contribute to understanding the molecular basis of self-renewal [3].

Stem Cell Aging

Aging involves a slow deterioration of tissue function, including an elimination of new growth and decreased capacity for repair. Aging is also associated with increased cancer incidence in all tissues that contain stem cells. These observations suggest a link between aging and stem cell function because stem cells drive growth and regeneration in most tissues, and because at least some cancers are thought to arise from the transformation of special stem cells—cancer stem cells [4]. It could be that much of age-related morbidity in mammals is determined by the influence of aging on stem cell function. Hematopoietic and nervous system stem cells were found to undergo strikingly conserved changes in their properties with aging [4]. Therefore, testing the hypothesis that there are conserved changes in gene expression within stem cells that regulate these age-related changes in function is of vital importance. It was hypothesized that stem cell aging is influenced by genes regulating the proliferative activity of stem cells during development, as well as by genes protecting stem cells from the wear and tear of adult life [5]. The identification of these genes might significantly contribute to the better understanding of the aging process.

Organogenesis from Stem Cells and Nature's Tissue Engineering Designs

The most fundamental question in organogenesis is: How do a small number of stem cells give rise to a complex three-dimensional tissue with different types of mature cells in different locations? The hematopoietic and nervous systems employ very different strategies for generating diversity from stem cells. The hematopoietic system assiduously avoids regional specialization by stem cells. Hematopoietic stem cells are distributed in different hematopoietic compartments throughout the body during fetal and adult life, and yet these spatially distinct stem cells do not exhibit intrinsic differences in the types of cells they generate [5]. This contrasts with the nervous system, where even small differences in position are associated with the acquisition of different fates by stem cells. While local environmental differences play an important role in this generation of "neural diversity," it was found that intrinsic differences between stem cells are also critical [6]. Part of the reason why different types of cells are generated in different regions of the nervous system is that intrinsically different types of stem cells are present in different regions of the nervous system. To understand the molecular basis for the regional patterning of neural stem cell function, scientists are studying how these differences are encoded.

An understanding of the mechanisms that regulate organogenesis from stem cells will make it possible to identify molecular links between stem cell function and disease. Scientists have combined gene expression profiling with reverse genetics and analyses of stem cell function in the hematopoietic and nervous systems to identify mechanisms that regulate organogenesis from stem cells and that lead to congenital disease when defective. For instance, Hirschsprung disease is a relatively common birth defect characterized by a failure to form enteric ganglia in the hindgut. It was found that it is caused by mutations in two pathways [the glial cell-derived neurotrophic factor (GDNF) and endothelin signaling pathways] that interact to regulate the generation and migration of neural crest stem cells in the gut. Mutations in these pathways lead to a failure to form the nervous system in the hindgut by preventing neural crest stem cells from migrating into the hindgut [7]. These insights raise the possibility of treating this disease with stem cell therapies that would bypass these defects.

References

1. Pavlovic M, Balint B (2006) The use of stem cells to repair cardiac tissue. Anest Reanim Transfuziol 34:129–150
2. Balint B, Pavlovic M (2006) Stem cells—biology, harvesting, extracorporeal purification and some aspects of their clinical application. Bilt Transf 52(2–3):2–10
3. Balint B, Stamatovic D, Todorovic M, Jevtic M, Ostojic G, Pavlovic M, Lojpur Z, Jocic M (2007) Stem cells in the arrangement of bone marrow repopulation and regenerative medicine. Vojnosanit Pregl 64(7):481–484 (VSP)
4. Balint B, Pavlovic M, Jevtic M, Hrvacevic R, Ignjatovic L, Mijuskovic Z, Blagojevic R, Trkuljic M (2007) Simple "Closed-cirquit" group–specific immunoadsorption system for ABO–Incompatible kidney transplants. Transfus Apher Sci 36(3):225–233
5. Pavlovic M (2008) VSELs concept (Review). MNE Medica, No 1, Year I, 16–17, ref. p.43
6. Pavlovic M, Todorovic M, Todorovic V, Tyagi V, Balint B (2009) Adult stem cell research and regenerative therapy in neurological diseases: limitations and perspectives Part I. Basic research data necessary for understanding stem cell treatment approaches in neuro-regenerative therapy. Anest Reanim Transfuziol 55(1–2):4–14
7. Pavlovic M, Todorovic M, Todorovic V, Tyagi V, Balint B (2009) Adult stem cell research and regenerative therapy in neurological diseases: limitations and perspectives Part II Neurological diseases and stem cell therapy. Anest Reanim Transfuziol 55(1–2):15–30

Chapter 9
Stem Cell Sources, Harvesting, and Clinical Use

*In theory, there is no difference between theory
and practice. But in practice, there is.*

– Yogi Berra

Hematopoiesis is an eventful and multifactorial continuous process in which a variety of blood cells are produced through proliferation and differentiation from a minor quantity of stem cells (SCs) [1]. A complex network of interactive matters and factors accomplishes the toti/pluri/multipotent SC survival, maturation, and multiplication. Namely, differentiation and proliferation of SCs in the bone marrow (BM) are regulated by the extracellular matrix and microenvironment provided by stromal cells [1]. These cells—including macrophages, fibroblasts, dendritic, endothelial, and other cells—stimulate hematopoietic SCs by producing growth factors like Flt3-ligand, stem cell factor or c-kit-ligand, various interleukins, granulocyte-macrophage colony-stimulating factor (GM-CSF), and granulocyte colony-stimulating factor (G-CSF). Other cytokines secreted by stromal cells regulate the adhesion molecules positioned on SCs, allowing them to remain in the BM or migrate to an area where the respective cell type is needed [1–4]. Thus, hematopoietic SCs could be defined as cells capable for self-renewal with high proliferative capacity and extensive potential to differentiate into all blood cells or numerous somatic cell types (the so-called SC plasticity due to "dedifferentiation" and/or "transdifferentiation") such as osteocytes, chondrocytes, hepatocytes, myocytes, cardiomyocytes, and even endothelial cells [5–9]. Thanks to these abilities, adult totipotent (having "unlimited" biological capacity), pluripotent, and multipotent SCs give rise to repopulation of recipient's BM with subsequent complete, stable, and long-term reconstitution of hematopoiesis known as engraftment. In addition, they are also capable of colonizing different tissues (homing) [9]. Thus, initial experimental and clinical studies have shown that "implantation" of autologous SCs into damaged and/or ischemic area induces their homing and subsequent "transdifferentiation" into the cell lineages of host organ, including collateral vessel

M. Pavlovic and B. Balint, *Stem Cells and Tissue Engineering,*
SpringerBriefs in Electrical and Computer Engineering,
DOI 10.1007/978-1-4614-5505-9_9, © The Author(s) 2013

formation. Angiogenetic growth factors (or genes encoding these proteins) promote the development of collateral arterioles, the process known as "therapeutic angiogenesis" or "neovascularization" [6–16].

The next part of this chapter has data related to practical aspects of SC harvesting, ex vivo manipulation, as well as therapeutic application (transplant) for treatment of patients with different hematological and other malignant or benign disorders. In a nutshell, SC transplant involves the administration of high-dose radio-chemotherapy and (re)infusion of collected cells to obtain abolition of disease, as well as to get hematopoietic reconstitution and general clinical improvement of patient's status. SC transplant with reduced-intensity conditioning (RIC) can be offered to patients who are ineligible for high-dose conditioning due to their age or co-morbidities [17–19]. Malignant hematological diseases have so far been the most common indication of this new treatment modality; it has been less often used for nonmalignant diseases. Nowadays BM and peripheral blood (PB) SC transplants are more common in adult allogeneic or autologous setting. Umbilical cord blood (UCB) transplants have provided hopeful results in pediatric settings, mainly when a matched unrelated SC donor is impossible to obtain [20–24].

Therefore, in practice for therapeutic use SCs can be collected by (a) multiple aspirations from BM; (b) cell harvesting by apheresis from peripheral blood after mobilization with chemotherapy and/or growth factors (rHuG-CSF); and (c) purification by processing from UCB. SCs collected from the stated sources can be clinically applied (transplanted) immediately following harvesting (allogeneic setting) or after a short-term storage in liquid state or a long-term storage in frozen condition—cryopreservation (autologous setting) [2, 5, 6, 25–29].

The Type and Influence of the Stem Cell Source

Bone marrow-derived stem cells. Historically, BM was the first source of SCs for transplant in experimental and clinical setting [30–35]. A marrow harvest is the same for an allogeneic donor as for an autologous patient. SCs are collected by multiple aspirations from the posterior and anterior iliac crest and (seldom) from sternum. The posterior iliac crest provides the richest source of marrow. The procedure is performed under sterile conditions in the operation room, while the donor is generally anesthetized. In order to provide required number of nucleated cells (TNCs), that is, $\geq 3 \times 10^8$/kg of body mass (kgbm), around 200 aspirations are required, where single aspirate volume is 2–5 mL. Immediately after the collection, cell aspirate should be filtered in order to remove bone and lipid tissue particles and/or cell aggregates. Anticoagulation is provided using solution containing citrate and by heparin diluted in saline (5,000 IU/500 mL), using autologous plasma or one of the cell culture medium for resuspension of collected cells [5, 6, 25, 26].

The target dose of collected marrow is 10–15 mL/kgbm. Thus, the volume of aspirate is relatively large (800–1,000 mL) and contains a high count of red blood cells. Accordingly, in order to prevent anemia in donors, blood for autologous transfusion should be collected around one week before SC collection and transplant [36].

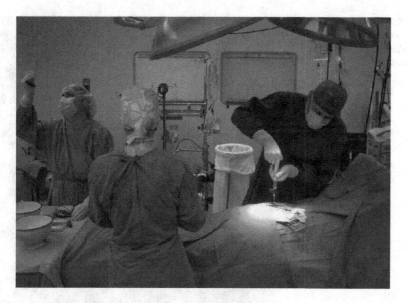

Fig. 9.1 Stem cell collection by multiple aspirations from bone marrow

Furthermore, as for aspirate volume, precise red blood cell number and/or plasma quantity reduction are required (by processing), especially for ABO-incompatible (major and/or minor) transplants or when cryopreservation is intended (autologous setting). Commonly used minimum target (after processing) of TNC count—for both autologous and allogeneic transplants—is 2×10^8/kgbm [26, 36] (Fig. 9.1).

The depletion of unwanted malignant cells or T-cells in final cell suspension is achieved by the ex vivo purging procedure using immunomagnetic cell selection [6]. These SC purification procedures (processing and purging or selection) enable reduction of the aspirate volume, i.e., reduction of red blood cell for around 80–90%, or even more precisely, the depletion of mentioned unwanted (malignant or T-cells) cells with efficacy $\geq$3–4 Log_{10} [3–6, 36]. Development of the ability to isolate selected and/or *ex vivo* expanded SCs in a large number is expected to broaden their beneficial therapeutic effects, since the limitation to many of SC applications has been the absolute number of defined target cells.

Peripheral blood-derived stem cells. CD34 is the cluster designation given to a transmembrane glycoprotein present on SC surface and some stromal cells. Cells expressing the CD34 antigen (obtained from BM or PB) are capable of complete reconstitution of hematopoiesis. The first SC harvests from PB were accomplished in "steady-state hematopoiesis"—but using several number (6 to 9) of cell harvestings and following cryopreservation was needed [32, 37]. Currently, SCs are harvested after mobilization by the use of chemotherapy and/or recombinant colony-stimulating factors (rHuG-CSF). The typical number of aphereses required is not more than one to three.

Briefly, SC harvesting from PB is performed by blood cell separators. The apheretic procedure is completely automated—uniform protocols are applied, with

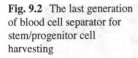

Fig. 9.2 The last generation
of blood cell separator for
stem/progenitor cell
harvesting

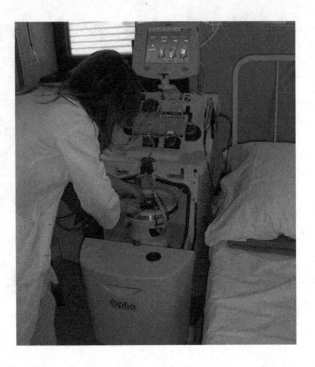

respect to the required standardization of the collected SC product. During apheresis, anticoagulation is achieved by ACD (solution A, USP). In allogeneic setting venous access is commonly realized through peripheral (ante-cubical) veins. However, for autologous donors—particularly for those who have received repeated chemotherapy—SC harvest should be performed across central venous access to circulation. Catheter-associated thrombosis is the most frequent possible complication. Also, there is around one percent hazard of the event of other complications associated with these catheters including infection, bleeding, or pneumothorax. Thrombocytopenia has been reported as a complication of rHuG-CSF administration and SC collection in healthy donors [5, 36].

Nowadays peripheral blood has largely replaced marrow as the source of SCs. The PB-derived SC transplant is characterized by (a) less invasive cell collection procedure; (b) absence of the risk of bone infections, general anesthesia, and work in the surgical division; (c) faster hematopoiesis reconstruction; (d) lower cell suspension volume (250–300 mL); and (e) absence or smaller level of contamination with tumor cells than in marrow aspirate. Clinical data comparing allogeneic peripheral blood vs. marrow transplants from HLA-identical siblings have also shown that peripheral blood-derived SC recipients have faster engraftment, improved hematopoietic and immune reconstitution, lower transplant-related morbidity, and a similar incidence of acute GvHD. Earlier studies reported a higher risk of chronic GvHD in allogeneic peripheral blood SC recipients. However, a recent prospective randomized study found no difference in the incidence of this complication [36–40] (Fig. 9.2).

Due to the reasons mentioned above, the number of patients treated by SCs harvested from PB is increasing worldwide, especially in autologous SC transplant setting. In order to obtain adequate SC yield, allogeneic donors are given rHuG–CSF 5–10 µg/kgbm per day. The count of CD34+ cells in the circulation begins to rise after the third day, and peaks following 5–6 days of rHuG–CSF administration. In autologous setting, patients are given higher rHuG–CSF dose (10–16 µg/kgbm daily) [6, 38–40]. In allogeneic setting, the first apheresis is performed typically on the fifth day of the rHuG–CSF application. The determination of optimal timing of an autologous SC harvesting is more complex and controversial. It is suggested as the optimal point to begin autologous SC collection when the leukocyte count exceeds 5–10×10^9/L. However, the leukocyte count commonly does not correlate with the CD34+ number. The count of circulating CD34+ cells evidently correlates with the optimized collection timing, as well as with the superior CD34+ yield in harvest. Generally, at peripheral blood CD34+ count $\cong 10$/µL, expected SC yield $\cdot 1 \times 10^6$/kgbm. It is also presented that for a CD34+ ≥ 20–40/µL of peripheral blood the possibility of the CD34+ $\geq 2.5 \times 10^6$/kgbm is 15% using one standard apheresis, and 60% after one large-volume SC harvesting [36, 38].

The target mononuclear cell count in harvest should be 314×10^8 per unit, that is, ≥ 2–4×10^8/kgbm, and CD34+ cells should be 330×10^6 per unit or ≥ 2–4×10^6/kgbm of the recipient in order to expect successful transplant. However, recent data support a benefit associated with greater CD34+ yield ($\geq 5 \times 10^6$/kgbm) compared to the minimum required cell quantity for engraftment ($\geq 1 \times 10^6$/kgbm) in autologous setting. Finally, results obtained in our SC transplant center confirmed that large-volume apheresis is efficient (CD34+ $\geq 5 \times 10^6$/kgbm) if the circulating CD34+ count was around 40–60/µL after mobilizing regiment [6, 38]. Although are generally accepted, the stated values of SC yields cannot guarantee stable and long-term BM repopulation after transplant. In order to achieve them, the following issues are needed: (a) quantity of processed blood ≥ 2–4 of patient's circulating blood volume per one apheresis (large-volume leukapheresis), that is, around 12–25 L for person with around 70–80 kgbm, and/or (b) apheretic procedures should be performed one to two (occasionally three) times on the average [5, 38–40] (Fig. 9.3).

Selection of CD34+ cells is associated with a reduction of count of tumor cells (autologous) or T-cells (allogeneic) in harvest. Since peripheral blood has become the major source of mobilized SCs for allogeneic transplant — but with high-level mononuclear cell contamination — effective technologies for unwanted cell depletion were developed to reduce the risk of disease recurrence or GvHD. The use of "positive" or "negative" immunomagnetic cell selection has been shown to be the most effective method to achieve ≥ 3–4 Log_{10} depletion of tumor or T-cells while preserving the high number of CD34+ cells in the harvest. Namely, the CD34+ cell recovery (CD34+ count in comparison with the number present before selection) should be at least 50–70% and cell purity (ratio of CD34+ vs. mononuclear cells) around 70–90% or more [5, 6, 36].

Patients who have been earlier treated with high-dose cytostatics may be poor responders to SC mobilization protocol with rHuG-CSF plus chemotherapy. The most efficient approach to obtain a sufficient SC yield in harvest from poorly

Fig. 9.3 The first PB harvest
obtained by MNC apheresis
in our center

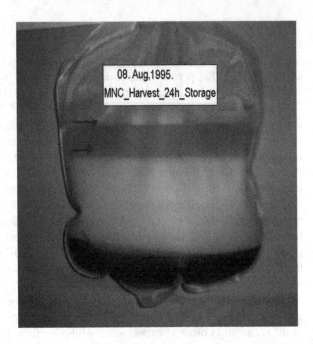

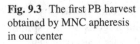

mobilized patients is undetermined still. Collection of SCs from BM together
with harvesting from PB has not improved the engraftment rate significantly.
Mobilization with rHuG-CSF alone is perhaps more effective than rHuG-CSF
plus chemotherapy for obtaining adequate CD34$^+$ yield [36]. Increased doses of
rHuG-CSF or polyethylene glycol-conjugated G-CSF (Pegfilgrastim), as well as
the use of rHuG-CSF plus IL-3 or CXCR4 inhibitor (Plerixafor) protocols has
also effectively mobilized some autologous donors (Fig. 9.4) [41–44].

Finally, application of the mobilization regimens and/or apheresis may result
in donor discomfort and several immediately (bone pain, agrypnia, and lassitude)
or delayed adverse events—but the side effects were rare and manageable
[36, 44–46].

UCB-derived stem cells. Patient's request for SCs only in ≤ 30% (related donor setting) and ≤ 85% (unrelated donor setting) can be solved finding an allogeneic HLA–
matched donor. Because of the limited availability of donors, attention has turned to
alternative sources of HLA-typed SCs. In recent years, UCB has emerged as a feasible alternative source of transplantable CD34$^+$ cells for allogeneic transplant,
mainly in patients who lack HLA-matched donors of BM- or PB-derived SCs
[20–25].

UCB is relatively rich in "more primitive" SCs that not only can be used to reconstitute the hematopoietic system but also have the potential to give rise to non-
hematopoietic cells (myocardial, neural, and endothelial cells, etc.) by transdifferentiation.

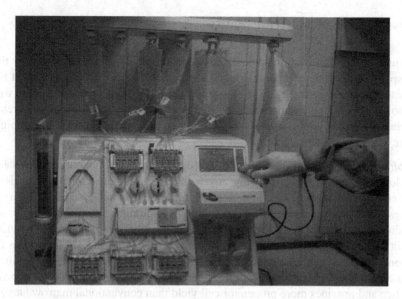

Fig. 9.4 Positive immunomagnetic CD34⁺ cell selection

The "naive" nature of UCB lymphocytes also permits the use of HLA-mismatched grafts at 1–2 loci without higher risk for severe GvHD relative to BM transplant from a full-matched unrelated donor. On the other hand, UCB is rich in primitive NK cells, which possess impressive proliferative and cytotoxic capacities and can induce Graft versus Leukemia (GvL) effect. The use of UCB is an accepted cell source for pediatric patients for whom smaller cell count is enough for engraftment, and for whom a matched unrelated allogeneic BM or peripheral blood SC donor is unavailable. However, a higher risk of graft failure was noticed in children weighing ≥45 kg. Since the number of SCs in UCB is limited and the collection can occur only in a single occasion — its use in adult patients can be more problematic [22–25, 36].

UCB volume is typically 100 mL (range 40–240 mL) with a TNC count $\approx 1 \times 10^9$ and CD34⁺ approximately 3×10^6 per unit. UCB can easily be cryopreserved, thus allowing for the establishment of HLA-typed SC banks. Because UCB-derived SC banking requires high financial investment and organizational efforts, banking efficiency should be optimized. An important determinant of banking efficiency is the ratio of collections that can be cryopreserved and supplied for transplant. Although there were reasons for removing UCB units that may be less amenable to improvement, such as low volumes and low cell counts, a number of obstetric factors influencing the outcome of collections could be evaluated further, including the time of cord clamping, length of gestation, length of labor, newborn's body weight, and weight of the placenta [24, 36].

Classification of SC Transplants

The use of high-dose radio-chemotherapy (conditioning regimen) followed by the transplant of allogeneic or autologous SCs is considered an effective treatment for hematological malignancies. In the last few years, the use of such a procedure was proposed also for non-hematopoietic neoplastic disorders (neuroblastoma, breast carcinoma, Wilms tumor, melanoma, lungs carcinoma) and also for some autoimmune disorders (sclerosis multiplex).

Autologous transplants. When is BM appropriate to use as a source of SCs for autologous resuscitation following myeloablative radio-chemotherapy or RIC depends on the marrow general state and/or infiltration with malignant cells. Fibrosis makes marrow not possible for SC collection by aspirations. Tumor cell infiltrates eliminate marrow as a transplant source as well. Prior pelvic irradiation, poor anesthesia risk, obesity, or patient refusal of marrow collection can limit marrow as an option. Mobilized autologous SCs from PB are commonly used in the above situations and in heavily treated patients. This procedure is nowadays in routine clinical practice and provides more progenitor cell yield than conventional marrow harvest and therefore earlier engraftment, that is, a faster hematopoietic recovery. Primarily for this reason, transplant of PB-derived SCs has practically replaced BM transplant in an autologous setting. As mentioned, autologous PB harvests involve mobilizing the SCs from the patient's BM compartment into the circulation using different growth factors, typically in combination with chemotherapy prior to collection. Once in the circulation, the SCs are collected by apheresis—conventional or large-volume leukapheresis [6, 38].

Allogeneic transplants. Transplant of allogeneic SCs is indicated in the treatment of patients with malignant disease—if they have HLA-matched donors. For patients with immunodeficiency, marrow failure, metabolism disorders, etc., the use of allogeneic SCs is imperative. However, there are also some atypical data related to treatment of severe aplastic anemia using autologous SCs [40]. Allogeneic transplant is associated with a risk that immunocompetent donor cells will react again recipient tissues (GvHD), despite immunosuppressive therapy administered. In adult "related allogeneic setting," the best results are obtained using completely HLA-matched (HLA-identical, i.e., six-antigen-matched donor/recipient pairs) transplants. There is a 25% chance of a sibling being a complete match, a 50% chance of a haplotype match, and a 25% chance of a complete mismatch. Pediatric patients are more tolerant of partially mismatched graft [1–4, 36].

Data obtained up till now has shown that the use of SC donor registers can successfully recruit unrelated donors for collection of BM- or PB-derived SCs. Thus, matched unrelated donor (MUD) searches can be initiated for approximately 70% of candidates without sibling donor. These protocols are superior because of high SC-grafting potential (allogeneic vs. autologous cells), following rapid hematopoietic reconstitution, and of GvL effect [36, 47]. For definitive choice, additional experimental and clinical trials for comparison of efficacy and outcome of autologous vs. allogeneic (related or unrelated) BM- vs. PB-derived SC transplant are required.

Syngeneic and haploidentical transplants. Finally, seldom recipient has an identical twin—a syngeneic transplant is optimal because the donor and recipient cells are genotypically identical (the first transplants performed in humans) [39]. On the other hand, syngeneic grafts do not induce graft vs. tumor, that is, GvL effect in recipient. Our knowledge of the immunobiology of SC transplant across major histocompatibility complex (MHC) barriers—haploidentical transplants—has increased significantly over the past decades. The key reason (or limitation) for realization of the potential of haploidentical SC transplant is the absence of an HLA-matched related donor in the majority of families. On the other hand, the conversion of a new hypothetical therapeutic option into the routine haploidentical SC transplant clinical practice is accepted and developed more slowly. The most critical complications of SC transplants across HLA barriers are the graft rejection and/or occurrence of GvHD. However, these adverse events maybe could be successfully prevented and treated using current pharmacologic approaches or manipulations during the haploidentical hematopoietic grafting process [1].

The Clinical Use of SCs—Conventional Transplant Setting

As mentioned, the earliest transplants in the treatment of patients were performed by infusion of SCs derived from the BM of identical twins (syngeneic transplant). The clinical use of SC transplants expanded with the application of cells harvested from either related or unrelated donors (allogeneic transplant) appropriately matched in the HLA-system, or even with a patient's own SCs (autologous transplant). Even though the phrase bone marrow transplant was initially applied, hematopoietic SC transplant could be more correct since SCs are required for rapid, complete, and long-term engraftment in clinical setting and—as it is described previously—their source may be not only BM but also PB or UCB [1, 36].

During the past decades, various diseases and disorders have been treated by SC transplants. Namely, different types of leukemia and additional hematologic malignancies, severe aplastic anemia and other BM failure states, immune deficiencies, some congenital disorders of metabolism, several solid cancers, as well as autoimmune disorders were the main indications for the therapeutic use of SCs. The indications for SC transplant can be classified as standard, clinical optional, developmental, or generally non-recommended (Table 9.1).

High-dose chemotherapy with allogeneic transplant or autologous SC support is a well-defined indication and standard approach in the treatment of presented malignancies and different benign diseases/disorders. The efficacy of SC transplant depends on the type of disease, its stage and sensitivity for radio-chemotherapy, age and general condition of the patient, and level of donor and recipient HLA-matching. However, an impressive advance in SC transplant outcome obtained when scientists

Table 9.1 Most common and relative indications for SC transplants

Hematological diseases/disorders
  Acute leukemia
    Acute myeloid leukemia
    Acute lymphoblastic leukemia
  Myelodisplasia
  Myeloproliferative disorders
    Chronic myeloid leukemia
    Other myeloproliferative disorders
  Chronic lymphoproliferative diseases
    Hodgkin lymphoma
    Non-Hodgkin lymphoma
      High-grade non-Hodgkin lymphoma
      Low-grade non-Hodgkin lymphoma
    Multiple myeloma
  Aplastic anemia
Other disorders
  Severe combined immune deficiency
  Acquired immune deficiencies
  Thalassemia
  Metabolic deficiencies
  Autoimmune disorders (e.g., sclerosis multiplex)
Nonhematological malignancies
  Neuroblastoma
  Wilms tumor
  Rhabdomyosarcoma
  Ewing sarcoma
  Breast, ovarian, testicular cancer

and clinicians recognized that SC transplants were most effective when used early in the disease course, rather than as a "last-option treatment" [1, 36].

Since 1996 European Group for Blood and Marrow Transplantation (EBMT) published special reports on the indications for the hematopoietic SC transplant in the current practice. Autologous and allogeneic SC transplants are now actual treatment options and have been incorporated in therapeutic algorithms for many diseases [48]. The EBMT recommendations are based on existing Clinical trials with guidance which must be considered together with disease risk factors, as well as the risk of the transplant procedure. Accordingly, conditioning regimens vary in their intensity and are classified as standard, RIC, or intensified.

As stated, there are standard, potential, and at present regularly non-suggested indications for SC transplants. Standard indications can be performed in any specialized center experienced in SC transplant procedures defined by EBMT and Joint Accreditation Committee EBMT-ISCT (JACIE) guidelines [49]. The clinical optional category is the most difficult one and should be performed in a special center with major experience in SC transplants. It encompasses many rare disorders and the paucity of data relating to transplant, variability in transplant procedures, and contribution of many patient risk factors such as co-morbidities. The third category incorporate developing SC transplants with little experience, like in some autoimmune diseases. Finally, generally non-recommended category includes SC transplant in early stages of disease when results of conventional treatment do not normally justify the additional risk of transplant-related mortality (TRM), or when the disorder is so advanced that the chance of success is too small (blast crisis of chronic myeloid leukemia (CML)).

Acute Leukemia

Acute myeloid leukemia (AML). AML is a heterogeneous clonal disease of hematopoietic progenitor cells which lose the ability for differentiation and for response to normal regulators of proliferation. Induction chemotherapy is necessary to induce complete remission (CR) in AML. Postinduction chemotherapy is aimed to eradicate/reduce residual disease. Combined chemotherapy induces CR in an average 60–80% of adult patients aged less than 60 years. After achieving remission, further chemotherapy is needed to prevent disease relapse. Postremission therapy for patients under the age 60 includes chemotherapy or SC transplant. Generally, SC transplant is performed after two or three cycles of chemotherapy. Patients are stratified into three main risk factors based on cytogenetics, molecular biology, and response to induction chemotherapy. About 40–50% adult AML patients have normal karyotype as intermediate risk factor. Unfavorable prognosis is associated with several entities including Wilms tumor gene 1, bcl-2, bax, the brain and acute leukemia cytoplasmic gene (BAALC), FMS-like tyrosine kinase-3 (FLT3), and mixed lineage leukemia gene (MLL). Some mutations of specific genes confer a more favorable prognosis, like mutations in the CCAT enhancer binding protein-α (C/EBP-α), CEBPA, and nucleophosmin NPM1 [50].

Concerning autologous SC transplant, better outcome is achieved when cells are harvested after the second or third course of chemotherapy. There are several advantages over allogeneic SC transplant due to unnecessary matched donor and lack of risk of complications such as graft versus host disease (GvHD). Also, immunological reconstitution is much faster and more complete. The major disadvantages are higher relapse rate due to lack of GvL effect and the contamination of graft with leukemic cells. In high-risk AML the role of autologous SC transplant is minimal, since the final clinical outcome is poor. The data of the Acute Leukemia Working Party (ALWP) registry indicate results as follows: leukemia-free survival (LFS) 43%, overall survival (OS) 51%, and response rate (RR) 53% [51].

Allogeneic SC transplant has had an extreme relevance in AML, since so many patients have been cured. In the last 10 years it has been shown that the results of MUD transplants as well as the results of cord blood transplantations are improved.

The age limit (>60 years) remains, but older patients are increasingly considered for a transplant and RIC protocols. Transplant procedures consist of several steps, first reducing the tumor burden by conditioning regimen (3–4log), immunosupression as the prevention of GvHD, as well as prevention therapy of infections due to neutropenia. The HLA barrier remains the most important factor, so the best results are those using an HLA-compatible family donor. HLA-compatible sibling transplant is standard reference for any other transplant. The results for AML patients transplanted in CR1 (first complete remission) give a probability of LFS from 45 to 70% [52].The relapse rate is approximately 20–25%. Additionally, MUD transplant represents the major effort of transplant centers in last decade. Now, there are more than ten million potential donors. The continuous improvement of HLA typing is associated with improved results. Current strategies in the treatment of AML patients suggest that for poor-risk patient, allogeneic SC transplant in CR1 is the only best solution. For intermediate-risk patients, who represent the majority of AML patients allogeneic SC transplant probably represents the best option if an HLA-matched sibling donor is available. Chemotherapy is not inferior to transplant in good-risk patients. Acute promyelocytic anemia is never recommended for transplant if the PCR for PML-RARα is negative, and chemotherapy is an optimal treatment option.

Acute lymphoblastic leukemia (ALL). Hematopoietic SC transplant has important role in the treatment for ALL patients. However, indications for SC transplant in first CR scheduling and procedures are still not completed yet. Most European ALL study groups define an indication for SC transplant in patients with unfavorable prognosis with survival probability less than 40% with chemotherapy alone [53]. Due to the poor outcome with intensive chemotherapy, SC transplant has always been the therapy of choice for Ph+ ALL. Nowadays the majority of these patients receive imatinib as front-line therapy, without an increase in TRM if SC transplant is performed thereafter. The survival after allogeneic SC transplant in first CR is 27–65% [54]. The presence of detectable minimal residual disease (MRD) beyond first consolidation chemotherapy provides an important new indication for SC transplant. There is a general agreement that all patients in second or later CR are candidates for SC transplant. This includes molecular relapse, which represents the reappearance of MRD above 10^{-4} to 10^{-3}. Recommendations of Germal Multicenter study Group for Adult ALL (GMALL) are given according to risk adapted approach. For high-risk patients allogeneic SC transplant (sibling or MUD) is recommended. In standard-risk patients, priority is allogeneic SC transplant in molecular nonresponders. Therefore, in patients with second CR, allogeneic SC transplant is necessary.

Myelodisplasia (MDS). MDS consists of a heterogenous group of clonal stem cell disorders. The spectrum of MDS varies from a disease with an indolent course over several years to a form with rapid progression to AML. Disease is characterized by deficiency of hemopoiesis associated with severe cytopenias, leading frequently to acute leukemia. The majority of MDS patients are older than 60 years. For this patient supportive care, including new biologic response modifiers is the mainstay of therapy. A therapeutic dilemma exists in MDS due to the multifactorial etiopathogenesis and various stages of the disease, as well as elderly age of patients at diagnosis. The therapeutic approach vary from low-intensity treatment plus supportive therapy to intensive chemotherapy combined with SC transplant. High-intensity therapies generally aim to alter the disease's natural history—improving survival, and decreasing progression to acute leukemia—and are mainly used for high-risk disease. However, unfortunately most patients with MDS are too old to be considered for intensive treatment, such as SC transplant [36]. In older age comorbidities are common, so the RIC regimens have been widely used in MDS [55]. In a multivariate analysis, the 3-year relapse rate was significantly increased after RIC, but the 3-year non-relapse mortality was decreased [56].

Myeloproliferative Disorders

CML. CML is a rare disorder with an incidence of 1–2 patients per 100,000 population each year. The translocation t(9;22) leads to chromosomal abnormality Ph chromosome that is present in more than 90% of this patients. Previously, the only curative option for CML patients was allogeneic SC transplant, and for those unsuitable for SC transplant, interferon (IFN)-α with or without cytosine-arabinoside. Imatinib as tyrosine kinase inhibitor (TKI) is much more superior than previous therapies in terms of hematological, cytogenetic, and molecular genetic responses [57]. However a significant minority of patients (15–20%) will fail to achieve complete cytogenetic responses, or proportion of 15–20% will lose their response [58]. The allogeneic SC transplant would be second-line therapy for that group of patients [59]. Also, there are the new generations of TKI like nilotinib or dasatinib used in terms of resistance on first generations of TKI [60]. The decision regarding the use of allogeneic SC transplant must now take into consideration prognostic factors like age, duration of CML, and nature of donor stem cells, but also the response to TKI. Within the EBMT recommendations, allogeneic SC transplant from sibling and well-matched donors remains a standard of care for the chronic and accelerated phases of CML, representing the only curative treatment for CML. Data relating to the incidence and risk of relapse after allografting were derived from qualitative RT-PCR for bcr-abl transcript. But, positive MRD for years post transplant without clinical progression may not always herald relapse. Donor lymphocyte infusion (DLI) has become the therapy of choice for patients who relapse after allogeneic SC transplant, especially for patients transplanted and relapsing in chronic phase [61]. Comparing the use of imatinib to DLI in this group of patients who relapsed after allogeneic SC transplant (31 pt), molecular remission is even better in the patients who received TKI [62].

Other myeloproliferative disorders. In other myeloproliferative disorders (MPD) including primary osteomyelofibrosis (OMF), myelofibrosis secondary to polycythemia vera (PV), and essential trombocythemia (ET), allogeneic SC transplant can be a therapeutical option, but depending on the risk factors and disease stage [63]. The allogeneic SC transplant should only be considered in patients with intermediate and high risk scores for OMF, since their expected median survival is less than 2 years. In patients with PV and ET occurrence of marrow fibrosis is the main criterion for allogeneic transplant. The use of RIC regimens resulted in less TRM and remision in 75% of the patients [64].

Chronic Lymphoproliferative Diseases

Hodgkin lymphoma. Newly diagnosed Hodgkin lymphoma (HL) although in advanced stage has good prognosis with standard chemotherapy. In patients with chemosensitive relapse or in second CR high-dose therapy (HDT) and autologous SC transplant is nowadays considered to be the therapy of choice [65]. Also, in primary refractory disease, autologous SC transplant is an option. Allogeneic SC transplant is still an experimental procedure for patients with relapsed or refractory HL [66].

High-grade non-Hodgkin lymphoma (NHL). The outcome of patients with high-grade NHL is unsatisfactory with standard approaches [67]. These patients may be candidates for front-line HDT and autologous SC transplant, but in first relapse. Approximately half of this patient can be cured with immunochemotherapy, usually R-CHOP. In younger patients with relapsed or refractory chemosensitive disease, HDT can be curative in a significant subset. But in the era of rituximab combining with chemotherapy, autologous and allogeneic SC transplant is optional. Other strategies to improve the outcome of the patients are currently being explored. In Diffuse large B cell lymphoma (DLBCL) the addition of rituximab to pre- and post-transplant therapy is promising [68]. The role of allogeneic SC transplant remains uncertain in high-grade NHL. The use of RIC can reduce TRM but evidence of significant graft versus lymphoma is lacking.

Low-grade NHL. Folicular lymphoma (FL) is by far the most common low-grade lymphoma, accounting for 20% of malignant lymphoma in adults, but 40% of all lymphomas diagnosed in Western Europe and the USA. Based on FL international prognostic index (FLIPI), therapeutic outcome on standard immunochemotherapy is different [69, 70]. The use of HDT with autologous SC transplant in the treatment of low-grade NHL has not yet been established. The rationale for considering transplantation is that disease is incurable using standard therapies, and young patients will die of their disease. Using consolidation with autologous SC transplant in first remission demonstrated no statistically significant benefit in favor of first-line autologous SC transplant in patients with FL, and it should be reserved for relapsed patients [70]. As concerning allogeneic SC transplant, it is not recommended in first CR due to high rate of TRM and long disease course. It can be used in relapsed patients with RIC regimens, consisting of alemtuzumab, fludarabine, and melphalan with short-course cyclosporin as GvHD prophylaxis.

Multiple myeloma. Multiple myeloma (MM) is a plasma cell disorder with an incidence of 4–5 cases per 100,000/year. It remains incurable with conventional therapies. New therapeutic agents like bortezomib as inhibitor of NFkappaB, immunomodulatory drugs, and angiogenesis inhibitors improved the disease course with an increased overall response rate ≥80% [71]. HDT followed by autologous SC transplant prolonged OS in comparison to standard therapy resulted in prolongation of the OS more than 10 years. This approach presents standard of care for younger patients with MM. Regarding tandem use of autologous SC transplant, its use will decrease, mainly due to the thalidomide maintenance [72]. In high-risk MM patients including cytogenetic [t(4; 14), t(14; 16), t(14; 20), del p53, del 13, hypodiplody, complex karyotype] or those with progressive disease during induction therapy, the use of novel agents as bortezomib improved the clinical course of disease. In high-risk patients, induction therapy followed by autologous SC transplant plus allo-RIC represents an adequate clinical approach.

Aplastic Anemia

Hypoplasia or aplasia of BM and extreme reduction of polymorphonuclear count in circulation are typical parameters for categorization of aplastic anemia (AA): very severe AA < 0.2×10^9/L, severe AA (sAA) 0.2–0.5×10^9/L, and mild AA > 0.5×10^9/L. Research data confirmed immune-mediated etiology of disease: (a) reduced hematopoietic progenitor cell quantity in BM; (b) elevated T cell count and interferon gamma (IFN-γ) production; and (c) altered mesenchymal cell function, expressed in a decreased inhibitory effect upon T cells [40, 73–75]. The standard therapeutic approaches for AA are the use of immunosuppressive therapy – IST (mainly for older individuals) or allogeneic SC transplant (primarily for younger patients). Usually BM is the source of allogeneic SCs and cells can be collected in the steady-state hematopoiesis or following BM-priming (BM-activation) using rHuGM-CSF or rHuG-CSF in low doses [1–4, 36, 37]. Syngeneic bone marrow transplant in patients with severe aplastic anemia is rare, and usually requires pre-transplant conditioning to provide engraftment. We report on a patient with hepatitis-associated sAA who was successfully treated with syngeneic PB-derived SC transplant after a series of infectious and bleeding complications. This is the second case of successful syngeneic SC transplant in hepatitis-associated sAA in the available literature [40].

The potential treatment by autologous SC transplant in sAA is still an innovative/pioneering therapeutic approach. Namely, majority of patients with sAA have residual hematopoietic SCs and could be successfully mobilized following initial IST and extended administration of rHuG-CSF [75]. The collected and cryopreserved cell could be afterwards clinically applied, that is, reinfused, for the treatment of sAA.

Finally, we have presented experience with sAA patient, successfully treated by autologous SC transplant [40]. The objective of our work was to optimize the mobilization/harvesting protocol/timing in order to obtain high CD34+ and especially a more primitive CD34+/CD90+ cell yield and recovery (using our own controlled-rate cryopreservation), with critical goal of improving conditions for complete and long-term hematopoietic reconstitution after autologous SC transplant. The results obtained in this study clearly confirmed that in sAA (with no allogeneic donor), autologous SC transplant can result in a complete and long-term medullar and hematological remission, as well as successful clinical outcome. To our best knowledge, this was also the second published case of autologous SC transplant in sAA [40].

Although SC transplant-related mortality and morbidity (TRMM) have decreased, SC transplants continue to pose multiple potential adverse effects and complications. The most frequent complications following SC transplant are even now engrafting failure, infections, and acute or chronic GvHD. Despite the advances made since the earliest days of transplant therapy, graft failure following allogeneic SC transplant is still a life-threatening complication. Antibiotic prophylaxis to prevent infections and supportive blood component care to minimize TRMM are critically important components in management for all, but especially for patients with chronic GvHD. To decrease GvHD incidence, cell harvest can undergo T-cell depletion and patients are treated prophylactically with a variety of immunosuppressive drugs. Less toxic transplants, in the form of RIC, are being actively investigated, with the promise of expanding indications for allogeneic transplants. That is, SC transplant with RIC can be offered to patients who are disqualified for high-dose conditioning because of their age or comorbidities. A careful proactive assessment to identify, treat, and, hopefully, prevent adverse events is essential to a successful SC transplant.

There is no doubt that SCs are considered an optimal targets for gene therapy. They are also ideal cells for gene transduction because of their ability to renew themselves and differentiate into progeny cells and generation of a self-perpetuating cell population that contains the transduced gene for the lifetime of the patient. Namely, specific diseases that could be candidates for gene therapy following gene transduction into SCs as vectors include thalassemia, sickle cell anemia, Fanconi anemia, purine nucleoside phosphorylase deficiency, chronic granulomatous disease, leukocyte adhesion deficiency, Gaucher's disease, and a variety of other metabolic deficiencies. UCB-derived SCs potentially could be used to correct genetic deficiencies at birth after successful gene transduction and autologous transplant.

References

1. Hoffman R (2005) Hematology basic principles and practice, 4th edn. Churchill Livingstone, New York
2. Balint B, Radovic M, Milenkovic L (1988) Bone marow transplantation. Vojnosanit Pregl 45:195–201
3. Ho AD, Hoffman R, Zanjani ED (2006) Stem cell transplantation. Biology, processing, and therapy. WILEY-VCH Verlag GmbH & Co KgaA, Weinheim

4. Bilko NM, Fehse B, Ostertag W, Stocking C, Zander AR (2008) Stem cells and their potential for clinical application. Springer, Dordrecht

5. Balint B (2006) Stem and progenitor cell harvesting, extracorporeal "graft engineering" and clinical use—initial expansion vs. current dillemas. Clin Appl Immunol 5(1):518–527

6. Balint B (2004) Stem cells—unselected or selected, unfrozen or cryopreserved: marrow repopulation capacity and plasticity potential in experimental and clinical settings. Maked Med Pregl 58(Suppl 63):22–24

7. Obradovic S, Rusovic S, Balint B, Ristic-AnĐelkov A, Romanovic R, Baskot B et al (2004) Autologous bone marrow-derived progenitor cell transplantation for myocardial regeneration after acute infarction. Vojnosanit Pregl 61(5):519–529

8. Pavlović M, Balint B (2006) The use of stem cells to repair cardiac tissue. Anest Reanim Transfuziol 34:129–150

9. Balint B, Stamatovic D, Todorovic M, Jevtic M, Ostojic G, Pavlovic M et al (2007) Stem cells in the arrangement of bone marrow repopulation and regenerative medicine. Vojnosanit Pregl 64(7):481–484

10. Makino S, Fukuda K, Miyoshi S, Konishi F, Kodama H, Pan J et al (1999) Cardiomyocites can be generated from stromal cells in vitro. J Clin Invest 103:697–705

11. Beltrami AP, Urbanek K, Kajstura J, Yan S–M, Finato N, Bussani R et al (2001) Evidence that human cardiac myocites divide after myocardial infarction. N Engl J Med 344(23):1750–1757

12. Shintani S, Murohara T, Ikeda H, Uenoi T, Honma T, Katoh A et al (2001) Mobilization of endothelial progenitor cells in patients with acute myocardial infarction. Circulation 103:2776–2779

13. Templin C, Luscher TF, Landmesser U (2011) Cell-based cardiovascular repair and regeneration in acute myocardial infarction and chronic ischemic cardiomyopathy—current status and future developments. Int J Dev Biol 55(4–5):407–417

14. Flynn A, O'Brien T (2011) Stem cell therapy for cardiac disease. Expert Opin Biol Ther 11(2):177–187

15. Forraz N, McGuckin CP (2011) The umbilical cord: a rich and ethical stem cell source to advance regenerative medicine. Cell Prolif 44(Suppl 1):60–69

16. Fukumitsu K, Yagi H, Soto–Gutierrez A (2011) Bioengineering in organ transplantation: targeting the liver. Transplant Proc 43(6):2137–2138

17. Craddock C, Bardy P, Kreiter S, Johnston R, Apperley J, Marks D et al (2000) Short Report: Engraftment of T-cell-depleted allogeneic hematopoietic stem cells using a reduced intensity conditioning regimen. Br J Haematol 111:797–800

18. Bethge WA, Faul C, Bornhäuser M, Stuhler G, Beelen DW, Lang P et al (2008) Haploidentical allogeneic hematopoietic cell transplantation in adults using CD3/CD19 depletion and reduced intensity conditioning: an update. Blood Cells Mol Dis 40(1):13–19

19. Robinson SP, Sureda A, Canals C, Russell N, Caballero D, Bacigalupo A et al (2009) Reduced intensity conditioning allogeneic stem cell transplantation for Hodgkin's lymphoma: identification of prognostic factors predicting outcome. Hematologica 94(2):230–238

20. Gluckman E, Broxmayer HE, Auerbach AD (1989) Hematopoetic reconstruction in a patient with Fanconi anemia by means of umbilical cord from HLA identical sibling. New Engl J Med 321:1174–1178

21. Rubinstein P, Dobrila L, Rosenfield RE (1995) Processing and cryopreservation of placental/umbilical cord blood for unrelated bone marrow reconstitution. Proc Natl Acad Sci U S A 92:10119–10122

22. Rogers I, Casper RF (2004) Umbilical cord blood stem cells. Best Pract Res Clin Obstet Gynaecol 18(6):893–908

23. Cohen Y, Nagler A (2004) Umbilical cord blood transplantation—how, when and for whom? Blood Rev 18(3):167–179

24. Skoric D, Balint B, Petakov M, Sindjic M, Rodic P (2007) Collection strategies and cryopreservation of umbilical cord blood. Transfus Med 17(2):107–113

25. Rowley SD (1994) Secondary processing, cryopreservation, and reinfusion of the collected product. In: Kessinger A, McMannis JD (eds) Practical considerations of apheresis in peripheral blood stem cell transplantation. Cobe BCT, Lakewood, pp 53–62
26. Rowley SD (1995) Standards for hematopoietic progenitor cell processing. In: Brecher ME, Lasky LC, Sacher RA, Issitt LA (eds) Hematopoietic progenitor cells: processing, standards and practice. AABB, Bethesda, pp 183–199
27. Balint B, Ivanovic Z, Petakov M et al (1999) The cryopreservation protocol optimal for progenitor recovery is not optimal for preservation of MRA. Bone Marrow Transplant 23:613–619
28. Balint B (2004) Coexistent cryopreservation strategies: microprocessor-restricted vs. uncontrolled-rate freezing of the "blood-derived" progenitors/cells. Blood Banking Transfus Med 2(2):62–67
29. Balint B (2000) From the initial efforts of cell freezing to the standardization of the blood cell cryopreservation. Bull Transfus 46:3–8
30. Thomas ED, Lochte HL (1959) Supralethal whole body irradiation and isologous marrow-transplantation in man. J Clin Invest 38:1709–1716
31. Mathe G, Jammet H (1959) Transfusions et greffes de moelle osseusse homologue chez des humaines irradies a haute danse accidentellement. Rev Fr Etud Clin Biol 4:226–238
32. Thomas ED, Ferrebee JW (1962) Prolonged storaged of marrow and its use in the treatment of radiation injury. Transfusion 2:115–117
33. Buckner CD, Stewart P, Clift RA (1978) Treatment of blastic transformation of CGL by chemotherapy, total body irradiation and infusion of cryopreserved autolognous marrow. Exp Hematol 6:96–100
34. Goldman JM, Th'ng KH, Park DS, Spiers ASD, Lowenthal RM, Ruutu T (1978) Collection, cyopreservation and subsequent viability of haemopoetic stem cells intended for treatment of chronic granulocytic leukemia in blast-cell transformation. Br J Haematol 40:185–195
35. Champlin R (1987) The role of bone marrow transplantation for nuclear accidents: implications of the Chernobyl disaster. Semin Hematol 24(2 Suppl):1–4
36. Balint B, Stamatovic D, Andric Z (2003) Stem and progenitor cell transplantation. In: Balint B (ed) Transfusion medicine. CTCI, Belgrade, pp 525–547
37. Stamatovic D, Balint B, Todoric–Zivanovic B, Marjanovic S, Lakic–Trajkovic Z, Malesevic M (2000) Second allogeneic bone marrow transplantation in treatment of patients with severe aplastic anemia following late graft rejection. Vojnosanit Pregl 57(5 Suppl):95–98
38. Balint B, Ljubenov M, Stamatović D, Todorović M, Pavlović M, Ostojić G et al (2008) Stem cell harvesting protocol research in autologous transplantation setting: large volume vs. conventional cytapheresis. Vojnosanit Preg 65(7):545–551
39. Savic A, Balint B, Urosevic I, Rajic N, Todorovic M, Percic I, Popovic S (2010) Syngeneic peripheral blood stem cell transplantation with immunosuppression for hepatitis-associated severe aplastic anemia. Turk J Hematol 27(4):294–298
40. Balint B, Stamatovic D, Todorovic M, Elez M, Vojvodic D, Pavlovic M, Cucuz–Jokic M M (2011) Autologous transplant in aplastic anemia: quantity of CD34$^+$/CD90$^+$ subset as the predictor of clinical outcome. Transfus Apher Sci 45(2):137–141
41. Wagstaff AJ (2009) Plerixafor: in patients with non-Hodgkin's lymphoma or multiple myeloma. Drugs 69(3):319–326
42. De Clercq E (2009) The AMD3100 story: the path to the discovery of a stem cell mobilizer (Mozobil). Biochem Pharmacol 77(11):1655–1664
43. MacVittie TJ, Farese AM, Smith WG, Baum CM, Burton E, McKearn JP (2000) Myelopoietin, an engineered chimeric IL-3 and G-CSF receptor agonist, stimulates multilineage hematopoietic recovery in a nonhuman primate model of radiation-induced myelosuppression. Blood 95:837–845
44. Nosari A, Cairoli R, Ciapanna D, Gargantini L, Intropido L, Baraté C et al (2006) Efficacy of single dose pegfilgrastim in enhancing the mobilization of CD34$^+$ peripheral blood stem cells in aggressive lymphoma patients treated with cisplatin-aracytin-containing regimens. Bone Marrow Transplant 38(6):413–416

45. Li P, Zhang GY, Zhu P, Wu BQ, Niu Q, Xie Y (2007) Peripheral blood stem cell mobilization with low dose rhG-CSF in 56 unrelated healthy donors. Zhongguo Shi Yan Xue Ye Xue Za Zhi 15(2):348–351 [Article in Chinese]

46. D'Souza A, Jaiyesimi I, Trainor L, Venuturumili P (2008) Granulocyte colony-stimulating factor administration: adverse events. Transfus Med Rev 22(4):280–290

47. Petakov M, Balint B, Bugarski D, Jovcic G, Stojanovic N, Vojvodic D et al (2000) Donor leukocyte infusion—the effect of mutual reactivity of donor's and recipietnt's peripheral blood mononuclear cell on hematopoietic progenitor cells growth. Vojnosanit Pregl 57(5 Suppl):89–93

48. Wingard JR, Gastineau DA, Leather HL et al (2009) Hematopoietic stem cell transplantation. A handbook for clinicians. AABB, Bethesda

49. Link H, Schmitz N, Gratwohl A, Goldman JM (1995) Standards for specialist units undertaking blood and marrow stem cell transplants—recommendations from the EBMT. Bone Marrow Transplant 16:733–763

50. Falini B, Nicoletti I, Martelli MF, Mecucci C (2007) Acute myeloid leukemia carrying cytoplasmic/mutated nucleophosmin (NPMc+ AML): biologic and clinical features. Blood 109(3):874–885

51. Breems DA, Löwenberg B (2007) Acute myeloid leukemia and the position of autologous stem cell transplantation. Semin Hematol 44:259–266

52. Frassoni F, Labopin M, Powles R, Mary JY, Arcese W, Bacigalupo A et al (2000) Effect of centre on outcome of bone-marrow transplantation for acute myeloid leukaemia. Acute Leukaemia Working Party of the European Group for blood and marrow transplantation. Lancet 355:1393–1398

53. Gökbuget N, Hoelzer D (2008) HSCT for acute lymphoblastic leukemia in adults. In: Apperly J, Carreras E, Gluckman E, Gratwohl A, Maszi T (eds) Haematopoietic stem cell transplantation. The EBMT handbook, 5th edn. ESH, Paris, pp 372–379

54. Chaidos A, Kanfer E, Apperley JF (2007) Risk assessment in haemotopoietic stem cell transplantation: disease and disease stage. Best Pract Res Clin Haematol 20:125–154

55. de Witte T, Sanz G (2008) HSCT for myelodysplasia in adults. In: Apperly J, Carreras E, Gluckman E, Gratwohl A, Maszi T (eds) Haematopoietic stem cell transplantation. The EBMT handbook, 5th edn. ESH, Paris, pp 380–387

56. Martino R, Iacobelli S, Brand R, Jansen T, van Biezen A, Finke J et al (2006) Retrospective comparison of reduced-intensity conditioning and conventional high-dose conditioning for allogeneic hematopoietic stem cell transplantation using HLA-identical sibling donors in myelodysplastic syndromes. Blood 108:836–846

57. Druker BJ, Guilhot F, O'Brien SG, Gathmann I, Kantarjian H, Gattermann N et al (2006) Five-year follow up of patients receiving imatinib for chronic myeloid leukemia. N Engl J Med 355:2408–2417

58. Apperley JF (2007) Mechanisms of resistence to imatinib in chronic myeloid leukemia. Lancet Oncol 8:1116–1128

59. Niederwieser D (2008) HSCT for chronic myeloid leukemia in leukaemia in adults. In: Apperly J, Carreras E, Gluckman E, Gratwohl A, Maszi T (eds) Haematopoietic stem cell transplantation. The EBMT handbook, 5th edn. ESH, Paris, pp 388–397

60. Signorovitch JE, Wu EQ, Betts KA, Parikh K, Kantor E, Guo A et al (2011) Comparative efficacy of nilotinib and dasatinib in newly diagnosed chronic myeloid leukemia: a matching-adjusted indirect comparison of randomized trials. Curr Med Res Opin 27(6):1263–1271

61. Guglielmi C, Arcese W, Brand R et al (2002) Donor lymphocyte infusion for relapsed chronic myelogenous leukemia: prognostic relevance of the initial cell dose. Blood 100:397–405

62. Weisser M, Tischer J, Schnittger S et al (2006) A comparation of donor lymphocyte infusions or imatinib mesylate for patients with chronic myelogenous leukemia who have relapsed after allogeneic stem cell transplantation. Hematologica 91:663–666

63. Cervantes F, Rovira M, Urbano–Ispizua A, Rozman M, Carreras E, Montserrat E (2000) Complete remission of idiopatic myelofibrosis following donor lymphocyte infusion after failure of allogeneic transplantation: demonstation of a graft-versus-myelofibrosis effect. Bone Marrow Transplant 26:697–699

64. Rondelli D, Barosi G, Bacigalupo A et al (2005) Allogeneic hematopoietic stem cell trasplan-
    tation with reduced intensity conditioning in intermediate-or high risk patients with
    myelofibrosis and myeloid metaplasia. Blood 105:4115–4119
65. Sureda A (2008) HSCT for Hodgkin's lymphoma in adults. In: Apperly J, Carreras E, Gluckman
    E, Gratwohl A, Maszi T (eds) Haematopoietic stem cell transplantation. The EBMT handbook,
    5th edn. ESH, Paris, pp 444–463
66. Sureda A, Robinson S, Canals C, Carella AM, Boogaerts MA, Caballero D et al (2008)
    Reduced-intensity conditioning compared with conventional allogeneic stem-cell transplanta-
    tion in relapsed or refractory Hodgkin's lymphoma: an analysis from the Lymphoma Working
    Party of the European Group for blood and marrow transplantation. J Clin Oncol
    26(3):455–462
67. Vose JM, Zhang MJ, Rowlings PA, Lazarus HM, Bolwell BJ, Freytes CO et al (2001)
    Autologous transplantation for diffuse aggressive non-Hodgkin's lymphoma in patients never
    achieving remission: a report from the Autologous Blood and Marrow Transplant Registry.
    Blood and marrow Transplant Registry. J Clin Oncol 19:406–413
68. Lazarus HM, Zhang MJ, Carreras J, Hayes–Lattin BM, Ataergin AS, Bitran JD et al (2010) A
    comparison of HLA-identical sibling allogeneic versus autologous transplantation for diffuse
    large B cell lymphoma: a report from the CIBMTR. Biol Blood Marrow Transplant
    16(1):35–45
69. Solal-Céligny P, Roy P, Colombat P, White J, Armitage JO, Arranz-Saez R et al (2004)
    Follicular lymphoma international prognostic index. Blood 104:1258–1265
70. Gribben JG (2008) HSCT for low-grade non-Hodgkin's lymphoma in adults. In: Apperly J,
    Carreras E, Gluckman E, Gratwohl A, Maszi T (eds) Haematopoietic stem cell transplantation.
    The EBMT handbook, 5th edn. ESH, Paris, pp 442–453
71. Blade J, Rosiñol L, Sureda A, Ribera JM, Díaz–Mediavilla J, García–Laraña J et al (2005)
    High-dose therapy intensification compared with continued standard chemotherapy in multi-
    ple myeloma patients responding to the initial chemotherapy: long-term results from a pro-
    spective randomized trial from the Spanish cooperative group PETHEMA. Blood
    106:3755–3759
72. Koreth J, Cutler CS, Djulbegovic B et al (2007) High dose therapy with single autologous
    transplantation versus chemotherapy for newly diagnosed multiple myeloma. A systemic reb-
    view and meta analysis of randomised controlled trials. Biol Blood Marrow Transplant
    13:183–196
73. Young N, Calado RT, Scheinberg P (2006) Current concepts in the pathophysiology and treat-
    ment of aplastic anemia. Blood 108:2509–2519
74. Champlin RE, Perez WS, Passweg JR, Klein JP, Camitta BM, Gluckman E et al (2007) Bone
    marrow transplantation for severe aplastic anemia: a randomized controlled study of condi-
    tioning regimens. Blood 109(10):4582–4585
75. Bacigalupo A, Piaggio G, Podesta M, Van Lint MT, Valbonesi M, Lercari G et al (1993)
    Collection of peripheral blood hematopoietic progenitors (PBHP) from patients with severe
    aplastic anemia (SAA) after prolonged administration of granulocyte colony-stimulating fac-
    tor. Blood 82:1410–1414

Chapter 10
The HLA and ABO Systems in the View of Stem Cell Transplant (HLA Typization: Choice of Donors)

> *Each problem that I solved became a rule which served afterwards to solve other problems.*
>
> – Rene Descartes

The MHC is a large and stable region mapped to the short arm of the chromosome 6, encoding genes that have a lot of functions in immune and nonimmune response. MHC includes regions for the well-known and the most polymorphic until now discovered gene system, the human leukocyte antigen (HLA) genes. The extreme polymorphism of the HLA system derives from the existence of multiple alleles at several loci. It is estimated that more than 100 million different phenotypes can result from all combinations of alleles in the HLA system [1–5]. HLA genes are autosomal with codominant expression, inherited regularly as a haplotype. Since progeny receives one chromosome (haplotype) from each parent, four combinations of haplotypes are possible in newborn. Seldom HLA genes show chromosome crossover and these recombinants are then transmitted as new haplotypes to the offspring. These genes contribute to the recognition of self and nonself, to the immune response to antigenic stimulus, and to coordination of cellular and humoral immunity. Discrimination of self from nonself is realized primarily through the interaction of T lymphocytes with peptide antigens (epitopes presented by antigen presented cells), but only when the T-cell receptor binds both an HLA molecule and specific antigenic peptide [4–6].

MHC genes are divided into three regions. The class I region encodes more than 18 HLA genes and pseudogenes. HLA-A, HLA-B, and HLA-C genes are responsible for the corresponding A, B, and C antigens. Other class I genes (e.g., HLA-E, HLA-F, HLA-G) encode different nonclassical HLA proteins with limited level of polymorphism and expression. Several class I genes express nonfunctional proteins or nonprotein antigens (such as lipids). Genes unable to express a protein product are termed pseudogenes. The most centromeric on chromosome 6 is the class II region, which contains the 17 known HLA genes and pseudogenes. In general, genetic organization of the MHC Class II region is more complex. HLA-DR, HLA-DQ, and HLA-DP gene clusters code production of correspondingly named proteins characterized as class II antigens—which contain two dissimilar (alpha and beta) chains. Numerous genes of the HLA-DQ and HLA-DP clusters are most likely pseudogenes. Placed between the class I and the class II region, in the central part of MHC is a cluster of genes, designated as class III, encoding various biologically active proteins, such as complement components (e.g., C2, C4A, and C4B), tumor necrosis factor-α (TNF-α), lymphotoxin-α, lymphotoxin-β, etc. These genes are involved in various inflammation events and various aspects of stress, inflammatory, or immune response [6–8].

The products of HLA system are glycoproteins placed on the surface of cell membranes and play a key role in antigen presentation. HLA class I molecules are found on the surface of most nucleated cells, such as lymphocytes, granulocytes, monocytes, platelets, and cells of solid tissues. On the contrary, mature red blood cells (RBCs) do not have HLA antigens, but erythroid progenitors contain them. Class II molecules have more limited cell distribution; they are expressed on B lymphocytes and cells of monocyte–macrophage lineage always, and on T lymphocytes and some other cells after appropriate stimulation [6].

The transplant of hematopoietic stem cells (SCs) is a potentially curative therapy for a variety of hematological and non-hematological diseases. The main biomedical interest in the HLA system originated right from transplant biology, as a result of organ and tissue transplant practice [9–13]. Matching of donor and recipient for HLA antigens is essential for the success of SC transplant [14–16]. Namely, HLA and ABO antigens have a central role in the long-term survival of the transplants. In contrast, recognition of differences in HLA antigens is probably the first step in the rejection of transplanted organs or tissues. HLA system has also important role in the pathogenesis of different adverse effects of the blood component therapy, involving immune-mediated platelet refractoriness, febrile nonhemolytic transfusion reactions (FNHTRs), transfusion-related acute lung injury (TRALI), and graft-versus-host disease (GvHD). Finally, in response to pregnancy, transfusion, or SC transplant, immunologically normal persons are more expected to develop antibodies again HLA antigens than to any other blood group system [6].

As mentioned, HLA testing is an integral part of cell, tissue, and organ transplant [12–14]. In relation to genetic, that is, HLA relationship between donor and recipient, transplants can be classified as autologous, allogeneic, and syngeneic. Autologous transplant is not strictly an authentic transplant, but rather a resuscitation of a patient with his or her own SCs after myeloablative radio-chemotherapy. In fact, SCs are reinfused to repopulate the lethally or sublethally damaged recipient's bone marrow. Allogeneic transplant includes application of cells from another person, that is, HLA-matched related or unrelated donor in order to rescue the patient following an intensive antitumor therapy. In the special case of an identical twin recipient and donor, such transplants are referred to as syngeneic [6]. The donor selection is made essentially based on an HLA-match, while donor–recipient RBC-incompatibility is considered secondarily. Differences inside HLA system between donor and recipient represent the most important barrier to SC transplant.

HLA compatibility is required for engraftment and to prevent GvHD, regardless of immunosuppressive treatment of recipient. Both the recipient and the potential donor have to be tested for HLA-A, -B, -C, -DR, and -DQ antigens. The aim is exact matching of the HLA-A, -B, and -DR antigens. DNA typing is carried out on samples from both the donor and recipient for best evaluation of HLA class I and II compatibility. Although HLA-identical sibling donors stay the best choice and the most frequent for SC transplant, there are expanding requests and use of unrelated donors. In addition, a cross-match of recipient serum against donor lymphocytes is required. The positive result of cross-match with unfractionated lymphocytes or T lymphocytes is a contraindication to transplant [13–15].

HLA typing for HLA-A and HLA-B has typically been dependent on serologic methods. However, GvHD and engraft failures are due to phenotypically matched unrelated donors with considerable differences in alloantigens not identified by serologic techniques. Due to the heterogeneity of the class I antigens, serologic techniques are in practice for HLA-A and HLA-B typing in many centers, but molecular typing is being performed with increasing frequency, particularly in unrelated transplants. Mismatching for a single class I or II alleles increases mortality rate. The class I HLA-C antigens have possibly a less important role in the T-cell immune response because of their reduced polymorphism and low expression on the cell surfaces. However, it is generally accepted that HLA-C antigens can be identified by cytotoxic T lymphocytes and NK cells, which may be associated with an increased hazard of graft failure. In contrast, the risk of GvHD is greater with HLA class II disparity. Testing for HLA-DR and HLA-DQ is regularly performed by molecular technology, i.e., DNA-based technique. This method gives better resolution (including subtypes of these alleles) than conventional serologic technique. Recent studies have demonstrated the importance of recipient HLA-DRB1 and HLA-DQB1 allele disparity in the development of GvHD [6].

Finally, the HLA null alleles are characterized by the absence of a serologically detectable gene product. Since serological HLA diagnostics are progressively replaced by DNA-based typing methods considering only small regions of the genes, null alleles may be misdiagnosed as normally expressed variants. The failure to identify an HLA null allele as a non-expressed variant in the SC transplant setting may result in an HLA mismatch that is highly likely to stimulate allogeneic T cells and to trigger GvHD. Since the occurrence of HLA null alleles is around 0.3% or even higher, a screening strategy for HLA null alleles should be implemented in the clinical laboratory. It may consist of the combination of serology and standard molecular typing techniques [17].

Umbilical cord blood (UCB) has recently been discovered as an alternative SC source for allogeneic cell therapy in both adults and pediatric patients with hematological malignancies and marrow failure syndromes. The relative ease of procurement, tolerance of 1–2 antigen HLA mismatch, and lower than anticipated risk of severe GvHD have made UCB an attractive alternative to marrow-derived SCs. Given that adults are larger than children, there was still limited enthusiasm for the use of UCB in adults. The use of reduced-intensity or non-myeloablative preparative regimens to allow engraftment of UCB broadens the scope of patients who may benefit from allogeneic immunotherapy, particularly the elderly and medically infirm patients with no matched sibling donor [6, 16].

Recipients of allogeneic SC transplant meet with the hazard of GvHD even when the donor is a sibling who shares the antigens of MHC. Thus, even the perfect HLA match does not correspond to the best possible genetic match between donors and recipients. In addition to the HLA complex, other genetic systems operate and affect the outcome of SC transplant. These include minor histocompatibility systems, as well as a series of functional polymorphisms in cytokines and chemokines and receptors genes [6, 7].

Among the events that have essential effect on the result of treatment with SCs (besides GvHD) the incidence and degree of infectious complications have an important role. Polymorphisms and/or malfunction of genes controlling the immune response to pathogens is a critical reason for vulnerability to infection after SC transplant. These include the HLA class I and class II alleles, Toll and TLR genes, etc. NK alloreactivity induced by HLA class I epitope mismatching (a common state in SC transplant) may also be encountered between the donor and the recipient leading to potentially damaging or advantageous events. Thus, a knowledge of the role of the most important genetic factors (MHC and non-MHC) will provide the rationale for a full matching in SC transplant settings [6, 18]. ABO-incompatibility is not a contraindication for SC transplant, but it is an important element for survival of transplanted cells and development of immediate post-transplant or delayed complications, such as anemia due to immune-mediated hemolysis. Thus, if marrow aspirate contains incompatible RBCs or plasma, this is an indication for the RBC-depletion or plasma volume reduction by processing [18]. The major RBC-incompatibility occurs when ABO or other clinically significant antibodies (e.g., Kell, Duffy antibodies) in the patient's blood react with donor RBCs. In these situations, RBCs should be removed from the aspirate. Techniques for RBC-depletion include HES sedimentation and the use of cell processor or separator. In addition to RBC-depletion, the majority of transplant programs remove antibodies (typically anti-A and/or anti-B) from the patient's circulation by plasma exchange.

In a minor ABO-incompatibility, donor's plasma contains antibodies against ABO antigens that are present on the recipient's RBCs. It usually does not cause considerable hemolysis, but clinically significant delayed hemolysis can occur in patients receiving minor ABO-incompatible SCs. Consequently, reduction of the plasma content in marrow aspirate is required. Plasma can be removed by the centrifugation of the marrow aspirate and removing the plasma layer with a plasma extractor [18].

ABO-incompatibility and HLA-sensitization (positive cross-match) are the two main barriers to organ transplants because of the allograft antibody-mediated rejection (AMR). Nowadays, kidney transplants across immunological barriers are possible due to existence of highly effective preconditioning protocols (regimens) [19–22]. Pretransplant regimens in ABO-incompatible kidney transplant setting include anti-CD20 infusion, therapeutic plasma exchange (TPE), extracorporeal immunoadsorption (ECIA), and standard triple immunosuppression: tacrolimus/mycophenolate–mofetil/steroid [19, 20]. The weak-point of "conventional" TPE is

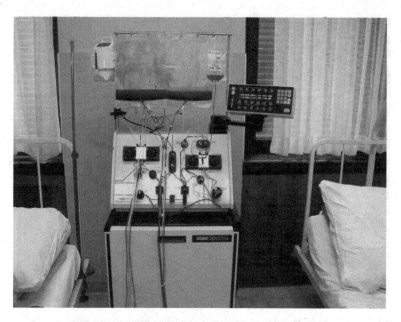

Fig. 10.1 Marrow aspirate processing after collection

the absence of depletion-selectivity. Contrary, selective TPE guarantees high blood purification with reduced risk of plasma constituent loss and virus transmission, since in this procedure the removed plasma by immunoadsorbed autologous plasma is substituted [20] (Fig. 10.1).

However, the optimal antibody titer in the recipient's circulation immediately prior to ABO-incompatible transplants is not yet determined. According to some authors [21] the acceptable titer is 16, while according to the others [22] it has to be ≤8. High-level antibody depletion by commercially available immunoadsorption columns (with synthetic group-specific oligosaccharides) is possible now, but the essential problem of these equipment is their excessive cost. TPE followed by donor-specific red blood cell transfusion (i.e., in vivo alloantibody binding) was firstly described in ABO-incompatible bone marrow transplant setting, although not without severe immune-mediated side effects [23]. The first techniques intended to in vitro depress the blood group-specific antibody reactivity were carried out in the early 80s of the twentieth century in our Institute too [24]. The principle of this method was the inhibition of antibodies with specific RBCs or soluble ABO and Lewis antigens from saliva (Fig. 10.2).

In summary, HLA antigens and antibodies play an important role in a number of transfusion-related events, especially in immunosuppressed patients. The rate of HLA-mediated platelet refractoriness among recipients of multiple transfusions is relatively high (more than 30%). Platelet refractoriness can be present when applied viable platelets are not successful in increasing recipient's platelet count. Besides HLA-mediated factors, refractoriness may be as a result of other causes such as

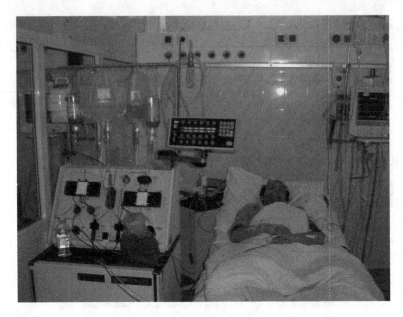

Fig. 10.2 ABO antibody depletion in recipient's circulation using plasma exchange

sepsis, disseminated intravascular coagulopathy, splenomegaly, and/or hyper-splenism or a combination of these. HLA antibodies, granulocyte- or platelet-specific antibodies, as well as proinflammatory cytokines (e.g., TNF, interleukins) are involved in the pathogenesis of the FNHTR. In TRALI acute HLA antibody and/or cytokine-mediated non-cardiogenic pulmonary edema develops, as a result of severe capillary leakage, in response to transfusion. The manifestation and evolution of GvHD depend on several factors such as the degree of HLA similarity between donor and recipient, the level of immunosuppression of recipient, and the quantity and viability of infused donor lymphocytes [6]. Conversely, donor-specific transfusions could result with positive effect on the overall outcome in patients treated by renal transplant [25, 26].

References

1. Snell GD, Dausset J, Nathenson S (1976) Histocompatibility. Academic, New York, pp 181–237
2. Bidwell JL, Navarrete C (eds) (2000) Histocompatibility testing. Imperial College Press, London
3. Howell MW, Navarrete C (1996) The HLA system: an Update and relevance to patient-donor matching strategies in clinical transplantation. Vox Sang 71:6–12
4. Tiercy JM (2002) Molecular basis of HLA polymorphism: implications in clinical transplantation. Transpl Immunol 9:173–180

5. Gruen JR, Weissman SM (1997) Evolving views of the major histocompatibility complex. Blood 90(11):4252–4265
6. Andric Z, Simonovic R (2003) The HLA system. In: Balint B (ed) Transfusion medicine. CTCI, Belgrade, pp 145–167
7. Roopenian DC, Simpson E (eds) (2000) Minor histocompatibility antigens: from the laboratory to the clinic. Landis Bioscience, Georgetown
8. Hviid TV, Hylenius S, Hoegh AM, Kruse C, Christiansen OB (2002) HLA-G polymorphisms in couples with recurrent sponaneous abortions. Tissue Antigens 60:122–132
9. Charron D (2005) Immunogenomics of hematopoietic stem cell transplantation. Pathol Biol (Paris) 53(3):171–173
10. Charron D, Suberbielle-Boissel C, Al-Daccak R (2009) Immunogenicity and allogenicity: a challenge of stem cell therapy. J Cardiovasc Transl Res 2(1):130–138
11. Yen BL, Chang CJ, Liu KJ, Chen YC, Hu HI, Bai CH, Yen ML (2009) Brief report-human embryonic stem cell-derived mesenchymal progenitors possess strong immunosuppressive effects toward natural killer cells as well as T lymphocytes. Stem Cells 27(2):451–456
12. Cecka JM (1997) The role of HLA in renal transplantation. Human Imunol 56:6–16
13. Moore SB (2001) Is HLA matching benefical in liver transplantation? Liver Transplant 7:774–776
14. Petersdorf EW (2004) HLA matching in allogeneic stem cell transplantation. Curr Opin Hematol 11(6):386–391
15. Müller CR (2002) Computer applications in the search for unrelated stem cell donors. Transpl Immunol 10:227–240
16. Koh LP (2004) Unrelated umbilical cord blood transplant in children and adults. Ann Acad Med Singapore 33(5):559–569
17. Elsner HA, Blasczyk R (2004) Immunogenetics of HLA null alleles: implications for blood stem cell transplant. Tissue Antigens 64(6):687–695
18. Balint B, Stamatovic D, Andric Z (2003) Stem and progenitor cell transplant. In: Balint B (ed) Transfusion medicine. CTCI, Belgrade, pp 525–547
19. Balint B, Pavlovic M, Jevtic M, Hrvacevic R, Ostojic G, Ignjatovic L, Mijuskovic Z, Blagojevic R, Trkuljic M (2007) Simple "closed-circuit" group-specific immunoadsorption system for ABO-incompatible kidney transplants. Transf Apher Sci 36:225–233
20. Balint B, Pavlovic M, Todorovic M, Jevtic M, Ristanovic E, Lj I (2010) The use of simplified *ex vivo* immunoadsorption and "multi-manner" apheresis in ABO/H-mismatched kidney transplant setting—phase II clinical study. Transf Apher Sci 43(2):141–148
21. Mazid MA, Kaplan M (1992) An improved affinity support and immunoadsorbent with a synthetic blood group oligosaccharide and polymer coating for hemoperfusion. J Appl Biomater 3(1):9–15
22. Shishido S, Asanuma H, Tajima E, Hoshinaga K, Ogawa O, Hasegawa A, et al (2002) ABO-incompatible living-donor kidney transplantation in children. Transplantation; 74(2):284–285
23. Lasky LC, Warkentin PI, Kersey JH, Ramsay NK, McGlave PB, McCullough J (1983) Hemotherapy in patients undergoing blood group incompatible bone marrow transplantation. Transfusion 23(4):277–285
24. Balint B, Radovic M, Taseski J, Calija B (1986) Investigation of the hydrosoluble type group-specific ABH and Lewis substances. Acta Antropol Yugoslav 23(1):5–12
25. Terasaki P (1989) Benefical effect of transfusion on kidney transplants. Vox Sang 57:158–160
26. Burlingam WJ, Solinger HW (1986) Action of donor-specific transfusions - analysis of three possible mechanisms. Transplant Proc 18:685–689

Chapter 11
Peritransplant Blood Component Therapy

I find that the harder I work, the more luck I seem to have.

– Thomas Jefferson

The clinical use of blood components (transfusion therapy) is an efficient method for the support of patients who underwent autologous or HLA-matched allogeneic hematopoietic SC transplant following chemotherapy (and nowadays just rarely total body irradiation—TBI) conditioning regimens. Thus, clinical qualification of blood replacement, as well as a variety of alternatives to "traditional" blood component support (autologous transfusions, blood substitutes, hematopoietic cytokines—growth factors), make the specific elements of current peritransplant transfusion therapy. The basic aim of transfusion therapy is the reconstitution of blood homeostasis through the improvement of red blood cell (RBC), platelet, white blood cell (WBC), or rarely coagulation factor deficiencies by replacement and/or stimulation of their production using cytokines. The events affecting the features of transfusion therapy are: (a) category and severity of patient's hematological deficit, and (b) type and quantity of blood component(s) or cytokine needed. These factors have to be determined before the initiation of blood replacement in all situations, and it is the highest priority in a high-quality transfusion therapy [1–3].

Despite the better donor selection, novel screening tests and procedures with the improved product quality, transfusion therapy is not administered without risks, and it occasionally results in a spectrum of adverse effects. Iatrogenic (transfusion-transmitted) infections and incompletely understood immunological adverse effects remain major concerns. Thus, clinicians should consider potential risks versus expected benefits of each transfusion. Only when the benefits clearly outweigh the risks should a blood transfusion be administered. Current standard and optimized transfusion practice—especially platelet support and hematopoietic stem and progenitor cell (SC) transplant—requires to obtain the highest possible cell yield, functional recovery, and the best therapeutic effect as well as the quality control to eliminate of hazardous technologies [4–7]. In this part of text indications or transfusion triggers

M. Pavlovic and B. Balint, *Stem Cells and Tissue Engineering*,
SpringerBriefs in Electrical and Computer Engineering,
DOI 10.1007/978-1-4614-5505-9_11, © The Author(s) 2013

for blood replacement in patients with depleted oxygen-carrying capacity and with hemostatic or immune-mediated disorders as well as adverse effects of transfusion therapy will be briefly described.

Clinical Use of Packed RBCs

Currently, when speaking in terms of blood transfusion, one mostly thinks of therapeutic use of packed RBCs. In view of transfusion therapy, anemia has to be characterized as a pathological condition with the decreased total RBC volume in which RBC transfusions are advantageous, or necessary [8].

Peritransplant donor support. The age and body mass of the donor and, in the autologous setting, the character of primary disorder as well as earlier myelosuppressive treatment influence not only on the SC yield, but also on the peritransplant-specific blood component requirements. Because of the large marrow volume (around 800–1,000 mL) collected, RBC transfusions may be required in both autologous and allogeneic settings. Commonly, allogeneic donors will have one autologous whole blood or RBC unit collected 1–2 weeks before SC harvest. Allogeneic RBCs for immunosuppressed autologous donors or children should be irradiated (with 25 Gy/unit) in order to prevent transfusion-associated GvHD in the recipient. Finally, in transplants that require marrow purification procedures, RBCs recovered (by processing) may be returned to the donor [1, 25].

Posttransplant blood component needs. Depending on the medical requests (e.g., allosensitization, repeated transfusions, organ and tissue transplant), different type of packed RBCs, such as washed, leukodepleted, or irradiated components are in therapeutic use. The application of each one has numerous potential advantages, but also has possible adverse effects. Determination of an appropriate threshold for initiating RBC replacement ("transfusion trigger") could help avoiding the unjustified blood transfusion, as well as "under-transfusion" situations, i.e., the inadequate support with RBCs [1, 8, 9].

Decision whether to transfuse packed RBCs or not, is usually based on the patient's hematocrit (Hct) and hemoglobin (Hb) values. However, rigid Hct and Hb thresholds—as universal transfusion triggers—have been considered inappropriate in several guidelines. Estimation of Hct and Hb does not always give a true description of RBC's lack because of the incorrect evaluation of total blood volume. Consequently, quantification of Hct and Hb should not be the only consideration in defining the transfusion trigger for RBC support [9–11]. Valid decision to transfuse should be made based on individual patient's general and cardiopulmonary status, tissue oxygen consumption rate, etc. The exclusively use of "rigid" transfusion triggers (Hct ≤0.30 or Hb ≤100 g/L) has little scientific confirmation and is outdated. After the discovery of AIDS and hepatitis C, the number of preferred RBC transfusions was clearly decreased—because the criteria for their performance became far more rigorous. Generally, RBC transfusion is rarely justified at Hb ≥ 100 g/L and is nearly always indicated when Hb ≤60 g/L [8, 10–13].

Table 11.1 Recommendations for prophylactic platelet transfusions

The threshold of 10×10^9/L is as safe as the higher levels for the majority of patients without the additional risk factors;

Risk factors include sepsis, usage of drugs such as antibiotics, and other abnormalities of hemostasis;

Platelet transfusions were also considered to be necessary at higher thresholds in immunosuppressed patients than in adults;

Higher threshold ($50–100 \times 10^9$/L) for platelet transfusions are needed to cover the invasive procedures (splenectomy, bone biopsy);

In acute DIC or massive blood transfusion with thrombocytopenia (platelet count $<50 \times 10^9$/L), platelet transfusions should be used in addition to coagulation factor replacement.

There are two main hazards of allogeneic RBC transfusions: the virus transmission and risk of acute immune-mediated hemolytic transfusion reaction (HTR). The majority of HTRs results from administration of ABO-incompatible RBCs. Sometimes RBC transfusion errors in practice lead to life-threatening and even fatal consequences. Many factors contribute to these errors, resulting from the misidentification of either the patient or the blood product. The HTR reaction could be promptly detected (careful monitoring of patient) and treatment quickly initiated [4–6, 14–16].

Platelet Transfusion Practice

Platelet concentrates (PCs) are suitable for prophylaxis or treatment of bleedings in patients with thrombocytopenia after SC transplant. Therapeutic use of PC is very effective, but the issue of the benefit of prophylactic platelet transfusion in prevention of hemorrhage remains controversial. A number of patients require specialized PCs, including leukodepleted (filtered), gamma-irradiated, or HLA-matched components [17].

Numerous reports, related to the determination of "platelet transfusion trigger", i.e., a threshold for their clinical use (Table 11.1) based on platelet count, have occurred recently [18–25]. Typically, severe hemorrhage (hematuria, hematemesis, and melena) is present at platelet counts $\leq 5 \times 10^9$/L. However, there are reports on cases of rare and moderate bleedings in patients with around 5×10^9/L platelets (due to the possible existence of a critical "individual thrombocytopenic bleeding threshold"). However, decision related to the introduction of PC transfusion should never be made exclusively upon the patient's platelet count [1]. Two types of platelet components are available in most hospital settings: platelet concentrates ("random-donor platelets") and apheresis-derived platelets ("single-donor platelets" or "pheresed platelets"). They differ in methods of preparation, platelet contents, and potential for adverse effects in the transfusion recipient. However, the hemostatic activity of platelets is similar in each component. At least, daily dose of platelets is approximately 0.5×10^{10} cells per kg of patient's body mass [8, 11, 25–27].

Each apheresis-derived PC is prepared from a single donor by platelet apheresis ($\geq 3-8 \times 10^{11}$ platelets in 150–200 mL plasma) [8, 17, 25].

PCs should be administered immediately after preparation, or after storage in liquid state up to 5–7 days on $20 \pm 2°C$, with permanent agitation of units using shakers [1, 25]. During storage the quantity and quality of PCs is progressively decreased. Long-term storage of platelets in frozen state requires the application of safe and effective cryopreservation procedures [19, 26]. Apheresis-derived PC contains 10^4–10^6 leukocytes (i.e., $\leq 1\%$ of leukocytes in a random-donor PCs) and less than 0.5 mL of contaminating RBCs, demonstrating the selectivity of the collection process. The low leukocyte content of apheresis-derived PC reduces the risk of leukocyte-mediated adverse transfusion effects [17].

Plasma and Coagulation Factor Replacement

Since the clinical use of fresh frozen plasma (FFP) is not free of adverse effects, data concerning plasma constituents and replacement effectiveness in clinical situations, such as posttransplant period are essential. Also, clinicians have to be informed about the alternative products for FFP in transfusion practice. The application of FFP should be reserved only for clinical situations in which plasma products have been proven evidently beneficial, or in conditions in which more specific replacement is not available [30]. Cryoprecipitate is a blood product containing factor VIII (FVIII:C), von Willebrand factor (vWF), factor XIII (FXIII), fibrinogen, and fibronectin [31, 32]. It is an insoluble cold precipitate formed when FFP is thawed between 1 and 6 °C. Each bag of cryoprecipitate contains between 150 and 250 mg of fibrinogen, 30–60 mg of fibronectin, and 80–120 units of FVIII. In addition, cryoprecipitate also contains approximately 40–70% (i.e., 80 units) of original vWF and approximately 30% (i.e., 40–60 units) of the initial concentration of FXIII [13, 16]. The recombinant coagulation factors allow rapid reversal of bleeding episodes, and therefore limit irreversible joint damage and other bleeding complications. In practice, for the termination of spontaneous bleedings, FVIII:C plasma activity should be improved up to the level of at least 0.30–0.60 IU/mL, and during the surgery from 0.50 to 1.0 IU/mL. Required quantities of FVIII:C can be calculated upon the volume of patient's circulating plasma as well as the initial and the desired FVIII:C plasma activity [25].

Granulocyte Support Practice

More recent advances in granulocyte collection, with the administration of rHuG-CSF to enhance granulocyte yield, have produced reestablished concern for this therapy (modern granulocyte transfusion support). Granulocyte collections are apheresis-derived. The use of rHuG-CSF has made possible markedly increased

Table 11.2 Transfusion therapy—the common adverse effects

Febrile non-hemolytic transfusion reaction (FNHTR);
Transfusion-induced immunomodulation (immunosuppression);
Hemolytic transfusion reaction (HTR);
Hypotensive reactions;
Transfusion-related acute lung injury (TRALI);
Platelet transfusion refractoriness;
Post-transfusion purpura;
Allergic and anaphylactic reactions;
Vasovagal syncope;
Circulatory overload;
Transmission of parasites and bacteria (transfusion-associated sepsis);
Complications of massive transfusion and apheresis;
Transfusion-induced hemosiderosis;
Transfusion-associated graft versus host disease (TA-GvHD);
Transfusion-transmitted (blood-borne) infections

blood granulocyte count of normal donors. Both higher granulocyte count in donor's circulation and processing of larger blood volume provide collection of evidently greater granulocyte doses (6–8×10^{10}/unit), compared with previously acceptable doses (1–2×10^{10}/unit). Consequently, renewed interest has extended the use of granulocyte transfusions in the treatment of transplanted adult patients in whom critical neutropenia (transfusion trigger ≤ 200–500/μL) was complicated by severe bacterial and/or fungal infections [1, 33].

Adverse Effects of Transfusion Therapy

Blood is a living tissue, and its transfusion from one individual to another is not free of hazards. Transfusion adverse effects (reactions or complications) might be immunologically or non-immunologically mediated, immediate or delayed, and may vary from mild to fatal. A variety of transfusion adverse effects are still a major concern in transfusion therapy (Table 11.2) [1, 34].

The most important hazards of transfusion are transfusion-transmitted infections and acute HTR caused by the use of ABO-incompatible RBCs. The advent of AIDS has raised concern regarding blood-borne diseases. Blood transfusion is safer than ever before through continual improvements in safe donor recruitment, i.e., screening (e.g., removal of high-risk donors), testing of each donation with a panel of viral markers, and appropriate clinical use of blood. The risk of residual infections is further reduced through inactivation of pathogens in blood components [35, 36]. As mentioned, RBC transfusions are rarely, if ever, justified in the treatment of patients with Hb approximately 100 g/L. If Hb concentrations range between 60 and 100 g/L, the occurrence and the intensity of symptoms and signs of inadequate tissue oxygenation will be essential for decision making concerning transfusion. For RBC administration, Hct and Hb values are practically of equal value, as well as the symptoms of inadequate oxygenation, together with the ability to compensate the blood loss, age of patients, and logistic aspects, such as the experience of the hospital physician, predictability, possible intensity, and duration of bleeding, etc. Exploration of a "common" or "universal" trigger for prophylactic platelet transfusions has to be continued because there are individual variations in bleeding tendency of patients with similar platelet counts, i.e., critical "individual thrombocytopenic bleeding threshold". It has been suggested that the optimal approach to prophylactic platelet support is to individualize it according to each patient's platelet count and clinical condition. However, this is only applicable in the centers with experienced medical staff to handle the orders for PC transfusions. Medical care should be initiated quickly, in order to avoid the expected sequels and complications of the basic disease.

References

1. Rebibo D, Morel P, Hauser L et al (2001) Blood transfusion surveillance: organization and results. Rev Prat 51:1332–6
2. Balint B (2004) Transfusion medicine. CTCI–Press, Belgrade
3. Wilkinson J, Wilkinson C (2001) Administration of blood transfusions to adults in general hospital settings: a review of the literature. J Clin Nurs 10:161–70
4. Bradbury M (2000) Cruickshank J.P. Blood transfusion: crucial steps in maintaining safe practice. Br J Nurs 9:134–8
5. Chiaroni J, Legrand D (2001) Immune safety in blood transfusions. Rev Prat 51:1311–7
6. Higgins C (2000) The risks associated with blood and blood product transfusion. Br J Nurs 9:2281–90
7. Moore BP (1995) Safer hemotherapy: the responsibilities of government, transfusion service, blood donors, and physician–users. Haematologia (Budap) 27:39–45
8. Balint B (2001) The "quality" (and quality control) of clinical indications to transfusion: an essential key to blood safety. In: Barbara J, Blagojevska M, Haracic M, Rossi U (eds) The future of blood safety, a challenge for the whole Europe: how can international regulations be implemented all over. Proceedings of the European school of transfusion medicine residential course, 2001, Oct 25–28, Sarajevo, Bosnia–Herzegovina. Milano: ESTM. p 117–24
9. Valeri CR, Crowley JP, Loscalzo J (1998) The red cell transfusion trigger: has a sin of commission now become a sin of omission? Transfusion 38:602–610
10. Capraro L, Nuutinen L, Myllylä G (2000) Transfusion thresholds in common elective surgical procedures in Finland. Vox Sang 78:96–100
11. Mollison PL, Engelfriet CP, Contreas M (1997) Blood transfusion in clinical medicine, 10th edn. Blacwell Scientific Publications, Oxford
12. Hillyer KL, Hillyer CD (2001) Packed red blood cells and related products. In: Hillyer KL, Strobl F, Hillyer CD et al (eds) Handbook of transfusion medicine. Academic, New York, pp 29–38
13. Högman CF (1999) Liquid–stored red blood cells for transfusion. Vox Sang 76:67–77
14. Bryan S (2002) Hemolytic transfusion reaction: safeguards for practice. J Perianesth Nurs 17:399–403
15. Balint B, Taseski J (2000) Therapeutic application of blood components and their recombinant alternatives in treatment of patients with hemostatic disorders. Vojnosanit Pregl 57(Suppl 5):69–80

16. Walter–Coleman S (1996) Transfusion therapy for patients critically ill with cancer. AACN Clin Issues 7:37–45
17. Roback JD, Hillyer CD (2001) Platelets and related products. In: Hillyer KL, Strobl F, Hillyer CD et al (eds) Handbook of transfusion medicine. Academic, New York, pp 53–61
18. Craig V, Bower JO (1997) Blood administration in perioperative settings. AORN J 66:133–43
19. Balint B, Vucetic D, Trajkovic–Lakic Z et al (2002) Quantitative, functional, morphological and ultrastructural recovery of platelets as predictor for cryopreservation efficacy. Hematologia 32(4):363–76
20. Balint B (2003) The function and therapeutic use of platelets and their alternatives. Vojnosanit Pregl 60:43–51
21. Murphy M (2000) Trigger factors for prophylactic platelet transfusions. In: Seghatchian J, Snyder EL, Krailadsiri P (eds) Platelet therapy. Current status and future trends. Elsevier, New York, pp 467–83
22. Consensus Conference on Platelet Transfusion (1998) Synopsis of background papers and consensus statement. Br J Hematol 101:609–17
23. Rebulla P (2000) Trigger for platelet transfusion. Vox Sang 78(Suppl 2):179–82
24. Wandt H, Frank M, Ehninger G et al (1998) Safety and cost effectiveness of a 10×10^9/L trigger for prophylactic platelet transfusions compared with the traditional 20×10^9/L trigger: a prospective comparative trial in 105 patients with acute myeloid leukaemia. Blood 91:3601–6
25. Brecher ME (2002) Technical manual, 14th edn. American Association of Blood Banks, Bethesda
26. Vucetic D, Balint B, Taseski J et al (2000) Quantification and investigation of morphology, ultra–structure and functional activity of platelets cryopreserved using six freezing protocols. Bullet Transfusiol 46:19–26
27. Balint B, Lj C, Taseski J et al (2002) The effect of leukoreduction on the proinflamatory cytokine concentration in platelet concentrates. Bull Transfusiol 48:6–15
28. Balint B, Radovic M, Paunovic D et al (1996) Examination of immune–inflammatory cytokines IL–6, IL–8 and TNF in platelet concentrates. Anest Reanim Transfusiol 25:39–43
29. Milenkovic L, Balint B, Skaro–Milic A et al (1995) Recovery and morphology of platelets prepared from buffy coat. Vojnosanit Pregl 52:18–24
30. Bucur SZ, Hillyer CD (2001) Fresh frozen plasma and related products. In: Hillyer KL, Strobl F, Hillyer CD et al (eds) Handbook of transfusion medicine. Academic, New York, pp 39–45
31. Bucur SZ, Hillyer CD (2001) Cryoprecipitate and related products. In: Hillyer KL, Strobl F, Hillyer CD et al (eds) Handbook of transfusion medicine. Academic, New York, pp 47–51
32. Balint B, Cernak I, Petakov M et al (2002) The use of single–donor fibrin glue prepared by recycled cryoprecipitation in experimental liver surgery. Haematologia (Budap) 32(2):135–46
33. Strauss RG (2003) Rebirth of granulocyte transfusions: should it involve pediatric oncology and transfusion reactions. In: Balint B (ed) Transfusion medicine. CTCI–Press, Belgrade, pp 807–48
34. Bharucha ZS (2001) Safe blood transfusion practices. Indian J Pediatr 68:127–31
35. Strauss RG (1999) Rebirth of granulocyte transfusions: should it involve pediatric oncology and transplant patients? J Pediatr Hematol Oncol 21:475–8
36. Higgins C (1994) Blood transfusion: risks and benefits. Br J Nurs 3:986–91

Chapter 12
Engraftment: Homing and Use of Genetic Markers

Do, or do not. There is no 'try'.

– Yoda

Homing refers to the stem cells' innate ability to travel to the right place in the body—the bone marrow—suited for making blood. The term "engraftment" means that the stem cells have begun their work; they are functioning properly within the marrow by producing various kinds of blood cells. Not only that bone marrow is recruited with fresh pool of concentrated stem cells, but it is also being gradually repopulated by those cells that emerge through differentiation of transplanted stem cells. Experimental evidence suggests that manipulated stem cells may lose some of their homing and engraftment abilities. If this evidence is true for humans as well, a troubling paradox may arise: The very success of an umbilical cord blood transplant could be undermined by the manipulations performed on stem cells—manipulations intended to increase their healing properties, not decrease or eliminate them. Research needs to clarify this. Work of this kind, at the University of Minnesota, is crucial to the success of stem cell expansion [1].

Genetic Markers

As in gene therapy, stem cells that are being genetically marked are activated to accept new genes. But instead of receiving genes that change their behavior, they receive genes that serve as "flags" or markers that are reproduced and expressed in every generation of subsequent cells. The markers can then be used by researchers to keep track of stem cell activity in the body after transplantation. For example, genetically marked stem cells are being used in a new experimental protocol at the University of Minnesota [1]. In this experiment, one-third of the stem cells the

M. Pavlovic and B. Balint, *Stem Cells and Tissue Engineering*,
SpringerBriefs in Electrical and Computer Engineering,
DOI 10.1007/978-1-4614-5505-9_12, © The Author(s) 2013

patient receives has been expanded and genetically marked outside the body in a lab [2]. Definitely, this could have a very significant impact on development of Tissue Engineering which is based upon the integration of stem cells, scaffolds, and active molecules that are enabling and facilitating the three-dimensional (3-D) growth of tissue cells.

References

1. Zwaka TP (2009) Use of genetically modified stem cells in experimental gene therapies. Stem Cell Information. NIH
2. Aubert J, Dunstan H, Chambers I, Smith A (2002) Functional gene screening in embryonic stem cells implicates Wnt antagonism in neural differentiation. Nat Biotechnol 20(12): 1240–1245

Chapter 13
Principles and Practice
of Stem Cell Cryopreservation

*"Problems worthy of attack prove their worth
by fighting back."*

– Paul Erdos

Recent extensive application of various cell-mediated therapeutic approaches has resulted in increased needs for both specific blood-derived cells and operating procedures to get minimized cell damages during their collection, processing, and storage in liquid or frozen state. The aim of cryoinvestigations is to minimize cell injuries during the freeze/thaw process (cryoinjury). Cryoinjuries may be the result of extensive cell dehydration and/or intracellular ice crystallization. The basic goal of cryopreservation is to maintain the cell viability—which may be defined as the ability of cells to perform their normal or near-normal function when transfused or transplanted. Generally, postthaw cell recovery is superior when the most appropriate freezing procedure and the best cryoprotective agent (cryoprotectant) are used. For blood progenitor or cell cryopreservation, glycerol, dimethyl sulfoxide (DMSO), and hydroxyetilstarch (HES) are regularly used, although in different concentrations. Despite the fact that cell freezing practice is already in routine use, some questions related to the optimal living cell cryopreservation are still unresolved.

Basic Cell Preservation Techniques

Generally, cell preservation methods could be categorized as: (a) a liquid-state storage at hypothermic temperatures, but $>0°C$ (usually $4±2°C$) for red blood cells and leukocytes or seldom for hematopoietic stem cells (SCs) up to 48 h; (b) a frozen-state storage ($-80°C$ to $-196°C$), the so-called cryopreservation [1].

In brief, cryobiology is a scientific discipline that estimates the effects of ultra-low temperature on cell integrity and functionality in the "refrigerated biological

M. Pavlovic and B. Balint, *Stem Cells and Tissue Engineering*,
SpringerBriefs in Electrical and Computer Engineering,
DOI 10.1007/978-1-4614-5505-9_13, © The Author(s) 2013

system" (cryobiosystem) and determines data/facts applicable in cryopractice. While the goal of the cell culturing is to keep cells in as near-normal conditions as possible by supplying nutrients essential for their metabolism, cooling/freezing is an effort to reduce the necessity of cells for energy. Precisely, after cooling, the need of cells for energy production (ATP synthesis) and its consumption (for protein synthesis, ion transport, and other biochemical activities) is decreased [2]. Cryopreservation is beneficial when cells appear to be biologically, chemically, or thermally unstable after liquid-state storage. Its primary purpose is to obtain both better cell recovery and well postthaw viability. Thus, cryopreservation includes specific approaches and techniques designed to extend "therapeutic shelf-life" of the cells (prolonged storage time) and to obtain minimum thermal damage (cryoinjury) [1]. The use of cryobiology for isolated cell preservation began in 1949 with the freezing of living animal sperm cells, using glycerol as a cryoprotectant [3]. Afterward, glycerol and DMSO techniques were applied for cryopreservation of different blood-derived cells [3–9]. The basic goal of these initial cryoinvestigations was to predict the cell response to freeze/thaw processes and cryoprotectant addition/removal. However, evaluation of cryobiological variables (biophysical, physicochemical, and other parameters responsible for cryoinjury) as well as standardization of practical aspects of cryopreservation is still a question of large interest to researchers and practitioners [10–17].

Although SC cryopreservation is now in routine use, certain freezing aspects should be revised to optimize specific cryobiosystem, i.e., to minimize the cryoinjury and maximize cell recovery [10–16]. Microprocessor-restricted (controlled-rate) freezing is a time-consuming process, which requires high-level technical expertise. Uncontrolled-rate ("dump-freeze" without programmed cooling rate) technique is less costly because it does not require a programmed freezing device. However, there are data [13, 18–23] that controlled-rate method is a high-class alternative to uncontrolled-rate technique [24–26] due to superior quantitative and functional cell recovery. Finally, for obtaining an effective cryopreservation, besides specific, i.e., optimized freezing method, the choice and use of appropriate cryoprotectant agent is required. At present, for SC and platelet freezing, DMSO and HES are commonly used as cryoprotectants, although in different concentrations [14–16, 22–33]. The next part of text recapitulates the current knowledge on cell cryoinjuries and effects of cryoprotectants as well as the conceptual and practical aspects of SC cryopreservation. In addition, recent activities in cryopractice, including our results obtained by the controlled-rate system (with compensated fusion heat) vs. uncontrolled-rate freezing ("dump-freezing" technique) in experimental and clinical settings will be summarized.

Cryoinjury: Its Origin and Mechanisms

The object of fundamental cryoinvestigations is to determine physicochemical and cryobiological attributes (including cell osmotic characteristics, water and cryoprotectant permeability coefficients) in the course of freezing, as well as to obtain better cell protection during cryopreservation. Cryoinjuries can be manifested as cell lesions, caused by the reduction of selected functions to the complete cell destruction and/or cytolysis [4–9]. Initially it has been considered that cryoinjuries are derived exclusively from the effect of extracellular ice crystals. In this sense, it has been proposed that the use of an adequately high freezing rate will help to avoid crystallization. However, a hypothetical cooling rapidity, so high that it would not cause cell cryoinjury, could not be achieved in practice due to thermo-dynamic limitations of heat transfer. In addition, during the "ultra-rapid" cooling, despite the expectations, complete cell destruction occurred [4–10]. At present it is considered that cryoinjuries result from the extensive volume reduction (cellular dehydration) and/or massive intracellular ice crystallization (mechanical damage) [6, 22, 31–33]. Although independent, these mechanisms can also act together. The first event is expressed primarily at low-rate ($\leq 10°$C/min) freezing and the second one in high-rate ($\geq 10°$C/min) freezing [10, 22]. In the course of low-rate freezing, extracellular crystallization occurs initially. These ice crystals do not cause mechanical cell damage (by membrane penetration) despite their physical presence [4–7, 10]. However, crystallization produces a progressive rise in osmotic pressure in extracellular area. Thus, extensive differences in the osmotic gradient (extracellular vs. intracellular) are formed, resulting in intensive intracellular water efflux. Consequently, cells become dehydrated (process known as "solution effect"). Dehydration and hypertonicity cause cell volume reduction, malforma-tions, and finally cytolysis [4–9, 20–22]. At high-rate (rapid) freezing, the osmotic gradient has no time to develop, and due to that dehydration and cell volume reduc-tion are minimal. However, intracellular ice crystal formation/growth and cell destruction have the most critical effect. The process is named "mechanical cell damage" [10, 22]. The quantity of "free" intracellular water can be increased by the release of "bound" water. This process is called "water desorption." The degree of mechanical cell damage is related to the total intracellular ice mass and the size of ice crystals [4–10]. Cryoinjuries may occur not only at freezing but also due to ice "recrystallization" and/or "dilution shock" during thawing. Small ice crystals will either form crystal "agglomerates" or they will amplify their mass by "recrys-tallization." During intracellular "recrystallization," mechanical cell damage occurs, while in the state of their extracellular occurrence cell dehydration will be expressed [4–10].

Inversely, rapid ice thawing causes a considerable increase in the extracellular water mass and a subsequent osmotic pressure decrease. Due to that, but also due to cryoprotectant slow efflux from cells, an extensive water influx into the cells ("dilution shock" or "cell swelling") will occur. Otherwise, cells will be more sen-sitive to the increase, than to the decrease of its volume, and thus, they will be destroyed [10, 22].

Therefore, determination of an optimal freezing rate (specific for each cell type and cryobiosystem) should be seriously considered. It is the speed of cooling which has to be high enough to prevent cell dehydration and adequately low to make efflux of water from the cell, possible. It would be ideal to find a cooling rate just less than

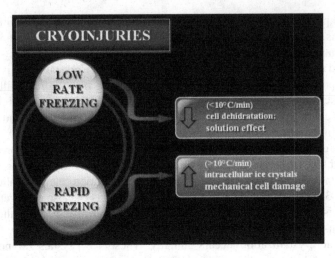

Fig. 13.1 Cryoinjuries: low-rate vs. rapid freezing procedures

the one which causes intracellular crystallization [10, 22]. Optimal freezing rate is the function of the ratio between cell surface and volume as well as of cell membrane permeability for water and its corresponding temperature coefficient—but it also depends on what type of cryopreservation strategy is applied [4–9]. Fig. 13.1.

Last but not the least, a higher degree of cell destruction has occurred when transition period from liquid to solid phase (fusion heat releasing) is prolonged. The released heat of fusion—if not considered during controlled-rate freezing—could result in additional temperature fluctuation. That is why the period of transformation from liquid to solid phase is prolonged, and its duration directly related to the degree of cryoinjury [5–12, 22]. Finally, in the course of rapid freezing, as stated, the quantity of "free" intracellular water can be increased as a result of "bound" water desorption. In accordance with this thermodynamic fact, more recently, a small second peak of heat release was detected between −30°C and −40°C [17]. We think that this "second heat release" might be partially caused by water desorption and additional crystallization. Consequently, there is potential need for revision of some cryodynamic parameters specific for controlled-rate cryobiosystems (Fig. 13.2).

The Use of Cryoprotectant Agents

Determination of the optimal freezing approach is essential, but it cannot solve all problems related to cell cryoinjury. Precisely, postthaw cell recovery and viability are high only when cryoprotectants are present in the cryobiosystem. They efficiently

Fig. 13.2 Device for controlled-rate freezing of stem cells

prevent or reduce the degree of cell thermal damages. As stated, the first applied cryoprotectant was glycerol [2]. Later on, numerous substances with high-level cryoprotective capacity were described [27–31]. One of them is DMSO which has demonstrated specific efficiency in the cryopreservation of SCs, platelets, lymphocytes, and granulocytes. DMSO is a transparent (colorless) fluid with a sulfur-like smell. It is a very polar molecule, which dissolves many water-soluble and lipid-soluble substances. DMSO has exothermic properties and should be mixed slowly with the cell suspension to dissipate the generated heat. Given intravenously (even in small concentrations) DMSO may cause certain unwanted effects such as nausea, vomiting, local vasospasm, etc. [22, 32–43]

Generally, cryoprotectants can be classified into nonpenetrating or extracellular and penetrating or intracellular compounds [29–31]. Mechanisms of their action are complex and only partially recognized. Due to the differences in its chemical and other properties, it is not possible to discover a cryoprotective mechanism common for all cryoprotectants. In a nutshell, extracellular agents could protect cells during high-rate freezing, reducing the intracellular ice crystal formation. However, intracellular cryoprotectants could provide protection in the course of low-rate freezing, decreasing the degree of cell dehydration [10, 22, 31]. Penetrating cryoprotectants increase intracellular solute concentration, and lower the temperature at which ice crystals form (reduced or "delayed" ice crystallization). Applied cryoprotectants, in higher concentration (but below the critical cytotoxicity) as well as reduced temperature, will increase intracellular viscosity. That is why the mobility of the water molecules and ice crystal formation and/or growth are slowed down. They decrease intracellular vs. extracellular osmotic gradient (reduced "solution effect") too [5–10]. Finally, certain cryoprotectants can change directly the cell membrane permeability for water, thus influencing the degree of cell dehydration [22, 29–33].

Cryoprotectants have other mechanisms of action, as well. The protective effect of penetrating cryoprotectants is obtained also due to their colligate effect, i.e., capacity for water binding [5–10]. Glycerol and DMSO are the first-class acceptors of hydrogen bonds; accordingly they can strongly bind a high amount of water and they have a high cryoprotective capacity. Thus, cells can be stored at below freezing temperatures without excessive intracellular ice crystal formation and extreme cellular dehydration [10, 29–33]. Although the cryoprotectant penetration (across the cell membrane) ability is critical for cell protection, it can be obtained by rapid (DMSO) and slow penetrating glycerol, as well as with nonpenetrating (HES) compounds. The temperature of cell is exposed to the cryoprotectant influence on the penetration rate also [22, 27–32].The main nonpenetrating cryoprotectant is HES, a compound initially used (together with DMSO) for granulocyte freezing. HES has a higher molecular weight than DMSO and predominantly, it acts extracellularly at the time of low-rate freezing. There are reports showing that it is possible to cryopreserve different blood cells with HES and DMSO using controlled-rate or uncontrolled-rate freezing. Cells frozen by these methods have adequate postthaw recovery and viability [14, 15, 24–26, 31–33].

In summary, cryoprotectants can express protective effect by the reduction of cell dehydration as well as by decreasing the intensity of intracellular crystallization. However, they cannot protect cells from already existing excessive dehydration or from the effect of previously formed intracellular ice crystals.

Stem Cell Cryopreservation Practice

The completion of SC transplant requires both efficient collection (by aspirations from bone marrow or apheresis from peripheral/cord blood) and cryopreservation techniques for obtaining an adequate cell yield and recovery. For SC cryopreservation, there are several well-known protocols, using primarily DMSO in autologous plasma, in various concentrations [11, 32–36]. Most authors suggest that the optimal cooling rate during cryopreservation is 1°C/min, while according to others it is 2–3°C/min [17, 36, 44, 45]. The transition from liquid to solid phase is also critical (because of releasing of the specific fusion heat), since a considerable reduction in cell viability has been observed when this period is prolonged. Thus, the optimal freezing rate for SCs has been shown to be 1°C/min, with an elevated cooling rapidity during liquid to solid phase transition (from 0°C to −10°C) period [11]. Finally, there is data showing that uncontrolled-rate freezing is also useful in SC cryopreservation [13–15, 24–26]. However, this technique produces an "unbalanced" or "changeable" cooling rate (around or over 3°C/min). In addition, the freezing bag or canister configuration as well as the volume of cell concentrate are critical parameters, which can radically modify the freezing velocity. In practice, bone marrow SC cryopreservation consists of the following steps: (a) aspirate processing (red blood cell and plasma reduction), equilibration (cell exposure to cryoprotectant), and freezing; (b) cell storage at −90±5°C (mechanical freezer), at temperature from

−120°C to −150°C (mechanical freezer or steam of nitrogen), or at −196°C (liquid nitrogen); and (c) cell thawing in a water bath at 37 ± 3°C.

The cryopreservation of peripheral blood vs. bone marrow-derived SCs has to be adapted to conditions which depend on the: (a) higher mononuclear cell count; (b) presence of plasma proteins; and (c) absence of lipid and bone particles in cell suspension [33]. Immediately after thawing in a water bath at 37 ± 3°C, cells are reinfused through a central venous catheter. Generally, patients tolerate the infusion of unprocessed SCs well, with no side effects [21, 32, 33, 42]. However, the grade of the potential reinfusion-related toxicity is associated with DMSO quantity in the cell concentrate. Thus, the recovery of SCs cryopreserved by 5% DMSO reported by Abrahamsen et al. is rather impressive [19, 46]. They showed that the use of 5% rather than 10% DMSO results in an improved CD34$^+$ cell recovery (low apoptotic and necrotic CD34$^+$ cell fraction) with a high potential for in vivo engraftment and ex vivo manipulations of these cells. Similar findings were also reported by Rowley et al. [15]. Finally, there are reports [14, 26] that DMSO concentrations lower than 5% (2.2 and 3.5% DMSO) are also sufficient for acceptable SC recovery.

Cryoprotectant agent can be removed by washing after thawing, but this procedure results in substantial SC loss [10, 22, 33]. The integrity of residual granulocytes is compromised within cryopreserved SCs and consequential DNA release during the thawing procedure may lead to cell "clumping" with resulting additional cell loss. To avoid this problem, a washing protocol by recombinant human deoxyribonuclease (rHu-DNase) is recommended [36]. The addition of rHu-DNase to cell concentrate seemingly proves to be effective in preventing "clumping" and it does not cause decreased expression of adhesion molecules, although it is not free of potential risks for patients. Moreover, the use of specific additives (e.g., membrane stabilizers) could improve postthaw cell recovery and it is probably a more effective approach than the decrease of DMSO concentration [34]. Our results are in agreement with the above-mentioned studies.

Namely, we have found that the recovery of mature population of the pluripotent and committed hematopoietic progenitors (CFU-Sd12 and CFU-GM) in the presence of 5% vs. 10% DMSO is superior [11, 20]. However, it has also been demonstrated that the recovery of very primitive pluripotent hematopoietic stem cells (Marrow Repopulating Ability—MRA) is better when 10% DMSO is used. These results imply a different "cryobiological request" of MRA cells in comparison with the nucleated cells and progenitors. Moreover, we have demonstrated that differences in cell recovery are not related to the changes in the total number of frozen/thawed cells, regardless of the use of cryopreservation strategy [11]. Our clinical studies showed that therapeutic use of the controlled-rate cryopreserved SCs in treatment of leukemia (ALL, ANLL, CML), multiple myeloma, Hodgkin's and non-Hodgkin's lymphoma, breast and ovarian cancer, and extragonadal non-seminal germ cell tumor resulted with high cell recovery (91%) and rapid posttransplant hematopoietic reconstitution—on the 11th day in average. Fig. 13.3.

Finally, in Transplant Center of MMA—in addition to the routine "initial" cell-mediated treatment using fresh ex vivo manipulated or unmanipulated cells—cryopreserved autologous SCs were applied for "repeated" therapy or "retreatment" of patients with large myocardial infarction [20, 40, 41, 47, 48]. Fig. 13.4.

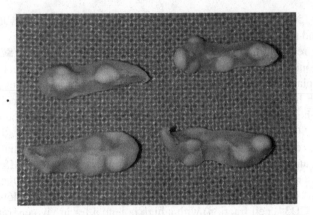

Fig. 13.3 CFU-Sd12 after controlled-rate cryopreservation using mice model

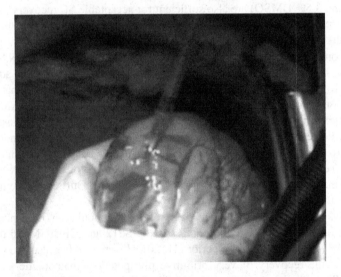

Fig. 13.4 BM-derived SC intramyocardial application in cardio surgery

By intensifying myeloablative therapy with SC rescue as well as by increasing the use of allogeneic transplants and different cell-mediated therapeutic approaches, the higher needs for both SCs and practical operating procedures appeared to minimize cell damage during collection and cryostorage. Cryobiosystem must effectively protect the biological and physical properties of cells that can be altered radically by the freezing/thawing process, and include techniques and materials appropriate for human use. Nowadays, a variety of different experimental and clinical freezing protocols are in practice, implying that the optimal procedure is not defined yet. Certain cryobiological techniques applicable in cryopractice should be revised for optimization of the postthaw cell recovery and viability.

References

1. Balint B, Radovic M (1993) Biophysical aspects of cryopreservation of hematopoietic stem and progenitor cells. Bull Transfus 21:3–8
2. Meryman HT (2007) Cryopreservation of living cells: principles and practice. Transfusion 47:935–45
3. Polge C, Smith AU, Parkes AS (1949) Revival of spermatozoa after vitrification and dehydration at low temperatures. Nature 164:666
4. Barnes DWH, Loutit JF (1955) The radiation recovery factor: preservation by the Polge-Smith-Parkers technique. J Natl Cancer Inst 15:901–906
5. Lovelock JE, Bishop MWH (1959) Prevention of freezing damage to living cells by dimethyl-sulfoxide. Nature 183:1394–5
6. Meryman HT (1956) Mechanics of freezing in living cells and tissues. Science 124:515–21
7. Mazur P (1966) Theoretical and experimental effects of cooling and warming velocity on the survival of frozen and thawed cells. Cryobiology 2:181–92
8. Lewis JP, Passovoy M, Trobaugh FE (1966) The transplantation efficiency of marrow cooled to −100°C at 2°C per minute. Cryobiology 3:47–52
9. Litvan GG (1972) Mechanism of cryoinjury in biological systems. Cryobiology 9:181–91
10. Balint B (2004) Coexistent cryopreservation strategies: microprocessor-restricted vs. uncontrolled-rate freezing of the "blood–derived" progenitors/cells. Blood Banking Transf Med 2(2):62–71
11. Balint B, Ivanovic Z, Petakov M et al (1999) The cryopreservation protocol optimal for progenitor recovery is not optimal for preservation of MRA. Bone Marrow Transplant 23:613–9
12. Balint B (2003) Blood cell cryopreservation. In: Balint B (ed) Transfusion Medicine. CTCI-Press, Belgrade, pp 479–95
13. Montanari M, Capelli D, Poloni A et al (2003) Long-term hematologic reconstitution after autologous peripheral blood progenitor cell transplantation: a comparison between controlled-rate freezing and uncontrolled-rate freezing at 80 degrees C. Transfusion 43:42–9
14. Lakota J, Fuchsberger P (1996) Autologous stem cell transplantation with stem cells preserved in the presence of 4.5 and 2.2% DMSO. Bone Marrow Transplant 18:262–3
15. Rowley SD, Feng Z, Chen L et al (2003) A randomized phase III clinical trial of autologous blood stem cell transplantation comparing cryopreservation using dimethylsulfoxide vs dimethylsulfoxide with hydroxyethylstarch. Bone Marrow Transplant 31:1043–51
16. Hubel A, Carlquist D, Clay M, McCullough J (2004) Liquid storage, shipment, and cryopreservation of cord blood. Transfusion 44:518–25
17. Liu JH, Ouyang XL, Lu LC, Gao D (2002) Thermometry of intracellular ice crystal formation in cryopreserved platelets. Zhongguo Shi Yan Xue Ye Xue Za Zhi 10(6):574–6
18. Rowe AW, Rinfret AP (1962) Controlled rate freezing of bone marrow. Blood 20:636
19. Abrahamsen JF, Bakken AM, Bruserud O (2002) Cryopreserving human peripheral blood progenitor cells with 5-percent rather than 10-percent DMSO results in less apoptosis and necrosis in CD34+ cells. Transfusion 42:1573–80
20. Balint B, Ivanovic Z, Petakov M et al (1999) Evaluation of cryopreserved murine and human hematopoietic stem and progenitor cells designated for transplantation. Vojnosanit Pregl 56:577–85
21. Balint B, Taseski J (2000) Cryopreservation of hematopoietic stem and progenitor cells. Maked Med Pregl 54:80–4
22. Balint B, Taseski J (1999) Long-term storage of blood cells by cryopreservation. Vojnosanit Pregl 56:157–66
23. Balint B, Stamatovic D, Todorovic M, Elez M, Vojvodic D, Pavlovic M, Cucuz-Jokic M (2011) Autologous transplant in aplastic anemia: quantity of CD34+/CD90+ subset as the predictor of clinical outcome. Transf Apher Sci 45(2):137–41
24. Stiff PJ, Koester AR, Weidner MK et al (1987) Autologous bone marrow transplantation using unfractionated cells cryopreserved in dimethylsulfoxide and hydroxyethyl starch without controlled-rate freezing. Blood 70:974–8

25. Clark J, Pati A, McCarthy D (1991) Successful cryopreservation of human bone marrow does not require a controlled-rate freezer. Bone Marrow Transplant 7:121–5
26. Halle P, Tournilhac O, TournilhacKnopinska-Posluszny W et al (2001) Uncontrolled-rate freezing and storage at –80 degrees C, with only 3.5-percent DMSO in cryoprotective solution for 109 autologous peripheral blood progenitor cell transplantations. Transfusion 41:667–73
27. Ahwood-Smith MJ (1961) Preservation of mouse bone marrow at –79°C with dimethyl sulfoxide. Nature 190: 4782, 1204–5
28. Rowe AW (1966) Biochemical aspects of cryoprotective agents in freezing and thawing. Cryobiology 3:12–8
29. Meryman HT (1971) Cryoprotective agents. Cryobiology 8:173–83
30. McGann LE (1978) Differing actions of penetrating and nonpenetrating cryoprotective agents. Cryobiology 15:382–90
31. Buchanan SS, Gross SA, Acker JP, Toner M, Carpenter JF, Pyatt DW (2004) Cryopreservation of stem cells using trehalose: evaluation of the method using a human hematopoietic cell line. Stem Cells Dev 13:295–305
32. Gorin NC (1986) Collection, manipulation and freezing of haemopoetic stem cells. Clin Haematol 15:19–48
33. Moroff G, Seetharaman S, Kurtz JW, Greco NJ, Mullen MD, Lane TA, Law P (2004) Retention of cellular properties of PBPCs following liquid storage and cryopreservation. Transfusion 44:245–52
34. Limaye LS, Kale VP (2001) Cryopreservation of human hematopoietic cells with membrane stabilizers and bioantioxidants as additives in the conventional freezing medium. J Hematother Stem Cell Res 10:709–18
35. Allan DS, Keeney M, Howson-Jan K et al (2002) Number of viable CD34+ cells reinfused predicts engraftment in autologous hematopoietic stem cell transplantation. Bone Marrow Transplant 29:967–72
36. Beck C, Nguyen XD, Kluter H, Eichler H (2003) Effect of recombinant human deoxyribonuclease on the expression of cell adhesion molecules of thawed and processed cord blood hematopoietic progenitors. Eur J Haematol 70:136–42
37. Beaujean F, Hartmann O, Kuentz M, Le Forister C, Divine M, Duedari N (1991) A simple efficient washing procedure for cryopreserved human hematopoietic stem cells prior to reinfusion. Bone Marrow Transplant 8:291–4
38. Makino S, Harada M, Akashik K, Taniguchi S, Shibuya T, Inaba S et al (1991) A simlified method for cryopreservacion of PBSC at –80°C without rate-controlled freezing. Bone Marrow Transplant 8:239–44
39. Halle P, Tournilhac O, Knopinska-Posluszny W et al (2001) Uncontrolled-rate freezing and storage at –80 degrees C, with only 3.5-percent DMSO in cryoprotective solution for 109 autologous peripheral blood progenitor cell transplantations. Transfusion 41:579–80
40. Marjanović S, Balint B, Stamatović D, Obradović S (1998) Early treatment of extragonadal non-seminal germ cell tumor with high-doses chemotherapy support with autologous hematopoietic progenitor cell transplantation. Arch Oncology 6:135–6
41. Stamatovic D, Balint B, Todoric B, Marjanovic S, Lakic-Trajkovc Z, Malesevic M (2000) Secondary allogeneic bone marrow transplantation in the patient with severe aplastic anemia following late graft rejection. Vojnosanit Pregl 57(5 Suppl):95–8
42. Sauer-Heilborn A, Kadidlo D, McCullough J (2004) Patient care during infusion of hematopoietic progenitor cells. Transfusion 44(6):907–16
43. Syme R, Bewick M, Stewart D, Porter K, Chadderton T, Gluck S (2004) The role of depletion of dimethyl sulfoxide before autografting: on hematologic recovery, side effects, and toxicity. Biol Blood Marrow Transplant 10:135–41
44. Balint B (2004) Stem cells—unselected or selected, unfrozen or cryopreserved: marrow repopulation capacity and plasticity potential in experimental and clinical settings. Macedonia Maked Med Pregl 58(Suppl 63):22–4
45. Spurr EE, Wiggins NE, Marsden KA, Lowenthal RM, Ragg SJ (2002) Cryopreserved human hematopoietic stem cells retain engraftment potential after extended (5–14 years) cryostorage. Cryobiology 44(3):210–7

46. Abrahamsen JF, Rusten L, Bakken AM, Bruserud O (2004) Better preservation of early hematopoietic progenitor cells when human peripheral blood progenitor cells are cryopreserved with 5 percent dimethylsulfoxide instead of 10 percent dimethylsulfoxide. Transfusion 44:785–9

47. Balint B, Stamatovic D, Todorovic M, Jevtic M, Ostojic G, Pavlovic M et al (2007) Stem cells in the arrangement of bone marrow repopulation and regenerative medicine. Vojnosanit Pregl 64(7):481–4

48. Balint B, Todorovic M, Jevtic M, Ostojic G, Ristanovic E, Vojvodic D et al (2009) The use of stem cells for marrow repopulation and in the field of regenerative medicine. Maked Med Pregl 63(Suppl 75):12–7

Chapter 14
Cord Blood Cell Cryopreservation

If you want to make an apple pie from scratch, you must first create the universe.

– Carl Sagan

Before the umbilical blood is frozen, it will first be introduced to a solution to help prevent it from being damaged while frozen. This solution is referred to as the cryopreservation solvent or cryoprotectant [1]. Once the blood has received this, it will begin to slowly freeze. Freezing it gradually is used as another preventative measure in guarding the cells against damage. Once the blood is frozen to a temperature of −96°C, it is transferred to a permanent storage freezer where it will remain frozen through the use of either liquid or vapor nitrogen. There are two different types of freezers commonly used in the preservation of cord blood stem cells. The first is the "BioArchive" freezer. This machine not only freezes the blood, but also inventories it and manages up to 3,626 blood bags [2]. It has a robotic arm that will retrieve the specified blood sample when required. This ensures that no other samples are disturbed or exposed to warmer temperatures. The second freezer system is the "Dewars" which is the most commonly used freezing system in the USA [2]. This unit is basically a well-insulated container with a lid. Because the lid needs to be opened every time a sample goes in or out of the freezer, there is a risk of the other samples being compromised. However, to help guard against this, temperatures on these freezer units are vigorously maintained and monitored. Cord blood cell samples are stored either in bags or in vials inside the freezers. Most cord blood banks will use only one type of storage since the bags or vials need to be evenly spaced on the racks inside the freezer. Therefore, the racks tend to be equipped to accommodate only one type of storage container. However, a few banks do use both bags and vials to store the cord blood. They have separate freezer units for each type of storage.

M. Pavlovic and B. Balint, *Stem Cells and Tissue Engineering*,
SpringerBriefs in Electrical and Computer Engineering,
DOI 10.1007/978-1-4614-5505-9_14, © The Author(s) 2013

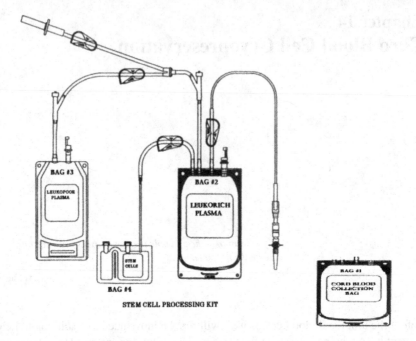

Fig. 14.1 Cord blood collection setting

A stem cells' viability in the freezer has yet to be definitely determined. Some scientists are conservative and put its freezer life at 10 years while some of them, 15 years. Many are expecting that the stem cells will be viable for up to 20 years, but studies have yet to conclusively determine this (Fig. 14.1).

References

1. Rubinstein P, Dobrila L, Rosenfield RE, Adamson JW, Migliaccio G, Migliaccio AR, Taylor PE, Stevens CE (1995) Processing and cryopreservation of placental/umbilical cord blood for unrelated bone marrow reconstitution. Proc Natl Acad Sci USA 92(22):10119–10122
2. Reboredo NM, Díaz A, Castro A, Villaescusa RG (2000) Collection, processing and cryopreservation of umbilical cord blood for unrelated transplantation. BMT 26(12):1263–1270

Chapter 15
Current Status and Perspectives in Stem Cell Research

*Perfection is achieved, not when there is nothing more to add,
but when there is nothing left to take away.*

– Antoine de Saint Exupery

Stem Cell Future

Research on stem cells is advancing knowledge about how an organism develops from a single cell and how healthy cells replace damaged cells in adult organisms. This promising area of science is also leading scientists to investigate the possibility of cell-based therapies to treat disease, which is often referred to as regenerative or reparative medicine. Therefore, in order to develop such treatments scientists are intensively studying the fundamental properties of stem cells, which include:

1. Determining precisely how stem cells remain unspecialized and self-renewing for many years.
2. Identifying the signals that cause stem cells to become specialized cells.

What Are the Unique Properties of All Stem Cells?

Scientists are trying to understand two fundamental properties of stem cells that relate to their long-term self-renewal:

1. Why embryonic stem cells can proliferate for a year or more in the laboratory without differentiating, but most adult stem cells cannot
2. What are the factors in living organisms that normally regulate stem cell and proliferation self-renewal?

M. Pavlovic and B. Balint, *Stem Cells and Tissue Engineering*,
SpringerBriefs in Electrical and Computer Engineering,
DOI 10.1007/978-1-4614-5505-9_15, © The Author(s) 2013

Discovering the answers to these questions may make it possible to understand how cell proliferation is regulated during normal embryonic development or during the abnormal cell division that leads to cancer. Such information would enable scientists to grow embryonic and adult stem cells more efficiently in the laboratory.

Regenerative Therapy in Myocardial Infarction: Mobilization, Repair, and Revascularization

Background and significance: Coronary heart disease is currently the principal cause of death in the United States. In 1997, 1.1 million Americans were diagnosed with acute myocardial infarction (AMI), and 800,000 patients underwent coronary revascularization. In patients with MI, scar tissue develops in the area of infarction resulting in a decrease in cardiac contractility. This damage is irreversible and can result in heart failure since cardiac cells cannot repair themselves.

There are a variety of cellular and molecular approaches to strengthening the damaged heart, focusing on strategies to replace dysfunctional, necrotic, or apoptotic cardiac cells with new ones of mesodermal origin [1]. A wide range of cell types such as myogenic cell lines, immortalized atrial cells, embryonic and adult cardiomyocytes, embryonic stem cells, teratoma cells, genetically altered fibroblasts, smooth muscle cells, bone marrow-derived cells, and adult skeletal myoblasts have all been proposed as useful cells in cardiac repair and may have the capacity to perform cardiac work [2–10]. Ultimately, it must be proven that cellular therapy aimed at cardiac repair not only improves pump function but also reduces mortality, morbidity, or both. During stem cell infusion due to organ damage with the goal to repair it, injury to a target organ is sensed by distant stem cells, which migrate to the site of damage and undergo alternate stem cell differentiation [11, 12]. These events promote structural and functional repair. This high degree of stem cell plasticity led researchers to investigate if dead myocardium could be restored by transplanting bone marrow (BM) cells. It was demonstrated that multipotent adult bone marrow hematopoietic stem cells (HSCs) and mesenchymal stem cells (MSCs) can repopulate infarcted rodent myocardium and differentiate into both cardiomyocytes and new blood vessels [11]. One of the first clinical studies done on the heart by Strauer et al. [28] reported that autologous intracoronary mononuclear bone marrow cell transplantation is safe and appears to improve cardiac function and myocardial perfusion in patients after acute MI ($n=10$) [12]. However, the authors concluded that further experimental studies, controlled prospective clinical trials, and variations of cell preparations are needed to determine the role of this new procedure for the treatment of patients after acute MI [12].

Stem Cells in Cardiac Repair and Revascularization (Engineering Damaged Heart Tissue)

Sources of cells for cardiac repair, and routes of their administration: Cells in current human trials include skeletal muscle myoblasts, unfractionated bone marrow, and circulating (endothelial) progenitor cells [13–15]. Cells in preclinical studies include bone marrow MSCs, multipotent cells from other sources, and novel progenitor or stem cells discovered in the adult myocardium [16–20]. Existing trials use intracoronary delivery routes (over-the-wire balloon catheters), intramuscular delivery via catheters (e.g., the NOGA system for electromechanical mapping), or direct injection during cardiac surgery. Not represented here is the theoretical potential for systemic delivery, suggested by the homing of some cell types to infarcted myocardium and strategies to mobilize endogenous cells from other tissue sites to the heart. Thus far, progenitor cells for cardiac repair have been delivered in three ways: via an intracoronary arterial route or by injection of the ventricular wall via a percutaneous endocardial or surgical epicardial approach [21–27].

The advantage of intracoronary infusion—using standard balloon catheters—is that cells can travel directly into myocardial regions where nutrient blood flow and oxygen supply are preserved, which hence ensures a favorable environment for cells' survival, a prerequisite for stable engraftment. Conversely, homing of intra-arterially applied progenitor cells requires migration out of the vessel into the surrounding tissue, so that unperfused regions of the myocardium are targeted far less efficiently, if at all. Moreover, whereas bone marrow-derived and blood-derived progenitor cells are known to extravasate and migrate to ischemic areas [28–31], skeletal myoblasts do not, and furthermore may even obstruct the microcirculation after intra-arterial administration, leading to embolic myocardial damage. By contrast, direct delivery of progenitor cells into scar tissue or areas of hibernating myocardium by catheter-based needle injection, direct injection during open-heart surgery and minimally invasive thoracoscopic procedures are not limited by cell uptake from the circulation or by embolic risk. An offsetting consideration is the risk of ventricular perforation, which may limit the use of direct needle injection into freshly infracted hearts. In addition, it is hard to envisage that progenitor cells injected into uniformly necrotic tissue—lacking the syncytium of live muscle cells that may furnish instructive signals and lacking blood flow for the delivery of oxygen and nutrients—would receive the necessary cues and environment to engraft and differentiate. Most cells, if injected directly, simply die [32]. For this reason, electromechanical mapping of viable but "hibernating" myocardium may be useful to pinpoint the preferred regions for injection [33]. Finally, in diffuse diseases such as dilated nonischemic cardiomyopathy, focal deposits of directly injected cells might be poorly matched to the underlying anatomy and physiology.

More complex and challenging is a series of pathobiological concerns, which have sent the scientific community from bedside to bench and back again. Certain patients' cells may be unsatisfactory, in their naive and unmanipulated state, which is now prompting systematic dissection of each step in progenitor cell function, from recruitment to plasticity. This task, in turn, is complicated by the fact that we do not yet understand the mechanisms underlying cell-based cardiac repair. For instance, the efficacy of skeletal muscle myoblasts [34] provided the impetus for human trials of skeletal muscle cells, but in the rabbit, diastolic functions are improved even by injected fibroblasts [35] and systolic performance improved to the same degree with bone marrow-derived cells as with skeletal muscle ones [36]. Along with the issue of skeletal muscle cells' electrical isolation from host myocardium, this prompts the question of how mechanical improvements arise even in this ostensibly straightforward instance. Another reason to consider potential indirect mechanisms is that studies have called into question the extent to which bone marrow-derived cells implanted in the heart form cardiomyocytes [35–37]. The majority of this review is, therefore, devoted to the biological horizons—namely mobilization, homing, neoangiogenesis, and cardiac differentiation, and to evolving new insights that may enable cell therapy for cardiac repair to surpass the present state of the art.

Mechanisms of action: Progenitor cells may improve functional recovery of infarcted or failing myocardium by various potential mechanisms, including direct or indirect improvement of neovascularization [38–43]. Paracrine factors released by progenitor cells may inhibit cardiac apoptosis, affect remodeling, or enhance endogenous repair (e.g., by tissue-resident progenitor cells). Differentiation into cardiomyocytes may contribute to cardiac regeneration. The extent to which these different mechanisms are active may critically depend on the cell type and setting, such as acute or chronic injury.

Hematopoietic or blood stem cells are critical to the daily production of over ten billion blood cells and are the basis for bone marrow transplant therapy for cancer, in the first instance. Rare and difficult to identify, these cells are extremely powerful at regenerating blood and immune cells but only if they travel to the proper location when introduced into the body. Typically the cells are infused into a vein, and they find their way to the bone marrow through a process that depends on largely unknown molecules. They are concentrated in bone marrow niche, the space close to internal wall of the bone and surrounded by macrophages and fibro, (Fig. 15.1) under very low oxygen tension compared to physiological.

The Molecules That Facilitate Homing of Stem Cells into Damaged Tissue Areas

How the stem cells may mediate this improvement in damaged tissue is not well understood. Ellis, Penn, and other investigators were, and still are considering additional mobilizing agents that may recruit a different subpopulation of stem cells, and are beginning to unravel the signaling cascade that directs stem cells to the infarct zone, as well as helps them transform or attract the other cells needed in the regenerative process [13]. Penn has identified the gene that codes for one of the

HSCs in the bone marrow niche
- hypoxic conditions within bone marrow - (mouse)

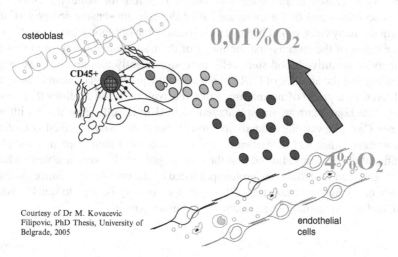

osteoblast

$0,01\%O_2$

CD45+

$4\%O_2$

Courtesy of Dr M. Kovacevic
Filipovic, PhD Thesis, University of
Belgrade, 2005

endothelial
cells

Fig. 15.1 HSCs in the bone marrow niche—hypoxic conditions within bone marrow (mouse). Courtesy of Dr. M. Kovacevic Filipovic, Ph.D. Thesis, University of Belgrade, 2005

signaling proteins (SDF-1) that helps guide stem cells to damaged heart tissue [13]. Unfortunately, after an MI, this gene is expressed for less than a week, and the homing signal it creates quickly fades. In his preclinical studies, Penn has found a way to turn this homing signal back on. As described in his later publication, he takes myoblasts from leg muscle, genetically engineers them to overexpress this homing protein, and then cultures those cells for 4 weeks until he has millions of them. Eight weeks after heart attack, those engineered cells are re-injected in controlled dosage with or without injections of the mobilizing agent, growth factor, granulocyte colony stimulating factor (G-CSF). Animals receiving such a treatment showed significant improvement in cardiac function (up to 90%) as well as neovascularization in the infarct zone. Besides SDF-1, Penn et al. have recently identified two other families of genes that seem to play a role in creating the homing signal of damaged myocardial tissue. Preclinical studies of these new gene candidates lie ahead. By better understanding how stem cells get recruited into the blood and directed towards damaged heart tissue, the researchers are beginning to find a way to extend the therapeutic window following MI, and more fully restore normal ventricular function.

As we already mentioned, within the bone marrow cavity, stem cells are usually found in the outer layer close to the inner surface of the bone (the space known as bone marrow niche, with a low oxygen supply). Since the process of remodeling bone takes place in the adjacent bone tissue and because studies by *Scadden's* group and others [23, 44, 45] have shown that bone-forming osteoblast cells are essential to the regulation of the stem cell environment, it seemed probable that fundamental interactions exist between the processes of bone formation and stem cell development. As increased extracellular calcium is required for bone formation, the researchers theorized that a molecule called the calcium-sensing receptor (CaR), present on many cells, might be key to the localization of blood stem cells [23, 44]. Examination of the spleens and the blood of the transgenic mice showed that the numbers of primitive blood stem cells were significantly elevated in those areas, indicating that the absence of CaR did not affect the production of stem cells by the fetal liver. In a group of normal mice that received radiation at doses that would destroy the bone marrow, transplantation of fetal liver cells from mice with and without CaR allowed the animals to survive, but those who received cells from CaR-negative mice had dramatically fewer stem cells in their bone marrow [44]. Additional experiments showed that the CaR-negative cells were unable to adhere to collagen I, an essential bone protein produced by the osteoblasts. "Since there are already drugs available that target this receptor, we may be able to quickly adapt these findings in animals to the treatment of human patients."

The Concept of VSEL Stem Cells and its Similarity with Other Pluripotent Stem Cell Candidates

Due to promising movement in the field of stem cell repair, as mentioned recently, by Ratajczak and colleagues [46] the concept of stem cell plasticity or trans-dedifferentiation created a high degree of hope and excitement. The supporters of stem cell plasticity postulated that stem cells isolated from easily accessible sources, such as bone marrow tissue, can be trans-dedifferentiated to stem cells for other organs (e.g., liver, pancreas, neural tissue, skeletal muscles, or heart). Thus HSC isolated from bone marrow could be employed for all types of tissue engineering (repair/regeneration). However, after the first optimistic reports, these promising results were not confirmed by other investigators. Actually, the concept of stem cell plasticity or trans-dedifferentiation created later on some disappointment, due to an assumption that HSC isolated from relatively easily accessible sources such as bone marrow (BM), mobilized peripheral blood, or cord blood, could be subsequently employed as precursors for other stem cells necessary for regeneration of various solid organs (e.g., heart, brain, liver, or pancreas). In all of these deliberations concerning stem cell plasticity [46–54] the concept that BM may contain heterogeneous populations of stem cells was surprisingly not taken carefully enough into consideration, until Ratajczak's group postulated that the regeneration studies that show the contribution of donor-derived HSC to tissues without excluding this possibility by not including the proper controls, could lead to the wrong interpretations. Thus, they stressed the argument that presence of *heterogeneous populations of stem cells in BM tissue* should be considered first; before experimental evidence is interpreted simply as trans-dedifferentiation/plasticity of HSC. They published the paper showing amazing evidence that BM stem cells are heterogeneous [48]. They proposed an alternative explanation of the plasticity of bone marrow-derived cells, by providing the evidence that bone marrow, in addition to HSC, contains an admixture of very rare tissue-committed stem/progenitor cells (TCSC) — for liver, pancreas, neural tissue, skeletal muscles, or heart — which in some experimental models gave a "false" impression that HSC changed their tissue commitment [47–54]. It is an important observation which questions the concept of stem cell plasticity [54–57] and points out that we should consider other sources of stem cells for tissue repair and regeneration (e.g., embryonic stem cells, amniotic fluid stem cells, and maybe the adult stem cells found in other already differentiated tissues such as adipose tissue, liver, olfactory tissue stem cells, etc.) [58–63]. This requires additional research since the pluripotent stem cells from these sources might be not as efficient in homing and repairing damaged tissue as they are coming from different environment.

This group has also confirmed, at both: the mRNA and protein level that BM contains, in addition to HSCs, a population of heterogeneous TCSC. Some of these cells possess internal markers characteristic of pluripotent stem cells and that they, similarly to HSC, could be mobilized from the BM into peripheral blood in situations of stress and, via the blood stream, may reach distinct organ locations to contribute to tissue repair/regeneration. So, the circulation of these TCSCs under steady-state condition plays an important physiological role in maintaining the pool of stem cells in distant part of the body. They might compete for SDF-1 — positive niches and this explains why is it possible to isolate HSCs from muscle or neural tissue and conversely muscle or neural progenitors from BM [53].

Finally, Kucia/Ratajczak et al. [52], described a new strategy to identify and isolate adult non-HSC from BM. They found that these cells, similarly to HSC, express CXCR4 and respond to an SDF-1 gradient; however, in contrast to HSC; these cells are CD45 negative, which is a fundamental difference between two categories. In summary, these very small embryonic like stem cells (VSELSCs) found for the first time in the bone marrow of mice, have very little cytoplasm with few mitochondria and internal markers of all three embryonic cell lineages, e.g., pluripotency (Oct4, Nanog, Rex-1 transcription factors). The morphology and phenotype of these mouse cells were similar to that found in the human MIAMI cells discovered by D' Ipolito, MAP (multipotent adult progenitor) cells (named also MAPSC) determined by Catherine Verfaillie and hBMSC detected by Youn Soop Yun in a latest, very condensed period of time, suggesting that it might be the same cell except that many investigators were not capable of reproducing them [64–66] (Figs. 15.2 and 15.3).

This is important in so much, since subpopulations of human mesenchymal stem cells (MSCs, hMSCs) exhibiting features of primitive adult pluripotent or multipotent stem cells have been described even earlier by different investigators. Cells with features of adult pluripotent stem cells have also been isolated from umbilical cord blood, peripheral blood, adipose tissue, but so far nobody gave such convincing data in distinguishing them from mesodermal stem cell, hematopoietic and pluripotent non-hematopoietic cell, neither they gave an undoubted reproducibility for pluripotent non-hematopoietic stem cell as it did Ratajczak's group, recently.

The reason for that was the most probably due to the fact that the superficial markers were sometimes at such a low concentration that the PCR amplification

Very Small Embryonic-Like Stem Cell (VSEL) non-hematopoetic pluripotent

M. Ratajczak et al., Leukemia 2006,20:857-869

- **VSELSC (MOUSE)**

  Only 0.02 % of all BMMNC
  Lower with ageing

  CD45-

- **Morphology**
  - High N:C ratio
  - Small (D=2–4 uM), posses large nuclei surrounded by a narrow rim of cytoplasm, and contain open-type chromatin (euchromatin) that is typical for embryonic stem cells

  - Scarce mitochondria
  - Embryonic-like bodies

- **Phenotype**
  - Sca+lin-CD45-

- **Internal markers**
  SSEA-1
  Oct-4
  Nanog
  Rex-1.

Fig. 15.2 An adult counterpart of embryonic stem cells present in mouse bone marrow

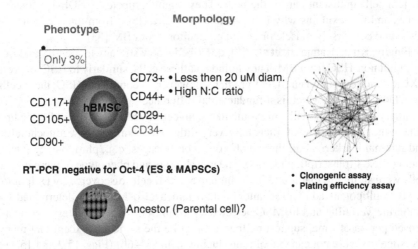

Phenotype

Morphology

Only 3%

CD117+   hBMSC

CD105+

CD90+

CD73+ • Less then 20 uM diam.
CD44+ • High N:C ratio
CD29+
CD34-

RT-PCR negative for Oct-4 (ES & MAPSCs)

Ancestor (Parental cell)?

- Clonogenic assay
- Plating efficiency assay

Fig. 15.3 Clonally expanded novel multipotent stem cells from human bone marrow regenerate myocardium after myocardial infarction

was required in order to detect them, the problem which was solved by Ratajczak's group. Their studies on FACS-purified BM cells combined with RT-PCR analysis revealed that human and murine pluripotent non-hematopoietic cells are CXR4+, CD34+, CD133+, CD45– and Sca-1+, lin–, CD45–, respectively. Thus, TCSCs which are already differentiated, and/or VSELSCs that are CD34+, CD133+ (human) and Sca-1+ (mice), in fact "contaminate preparations of bone marrow cells that are thought to be enriched in HSCs, only" (Fig. 15.3). This may explain why purified BM HSCs were found in several experimental settings to be "plastic" and able to "transdifferentiate" into various tissues as it was mentioned, although the controversies about plasticity of hHSCs always existed [36, 37, 67–69]. This fundamental discovery of a very deep pluripotent ancestor, with clearly defined and expressed markers, and a final proven distinction between hematopoietic and non-hematopoietic pool of adult pluripotent human stem cells, will finally improve the chances for more successful therapeutical intervention. This will greatly help to solve the problem of the source of pluripotent adult human cells for stem cell therapy with special reference to MI and other targeted diseases such as stroke, Parkinson's disease, spinal injury, muscular dystrophy, osteogenesis imperfecta and many others. The cells could be then collected by mobilization (using G-CSF as mobilizer) and apheresis and chosen fractions infused, when necessary, in order to repair targeted: infarcted (or other) tissue(s).

Heart failure—a severe deficiency in ventricular pump function—arises through a finite number of terminal effector mechanisms, regardless of the cause. These include: defects intrinsic to cardiac muscle cells' contractility due to altered expression or operation of calcium-cycling proteins, components of the sarcomere, and enzymes for cardiac energy production; defects extrinsic to cardiac muscle cells, such as interstitial fibrosis, affecting organ-level compliance; and myocyte loss, unmatched by myocyte replacement. Cardiac regeneration is robust for certain organisms such as the newt and zebrafish, in which total replacement can transpire even for an amputated limb, fin, or tail, via production of an undifferentiated cell mass called the blastema [71]. Such a degree of restorative growth might also be dependent on the retention of proliferative potential in a subset of adult cardiomyocytes [72] and is impossible in mammals under normal, unassisted biological circumstances. Several complementary strategies can be foreseen as potentially aiding this process: overriding cell-cycle checkpoints that constrain the reactive proliferation of ventricular myocytes [73]; supplementing the cytoprotective mechanisms that occur naturally, or inhibiting pro-death pathways [73]; supplementing the angiogenic mechanisms that occur naturally using defined growth factors or vessel-forming cells; or providing exogenous cells as a surrogate or precursor for cardiac muscle itself [74]. Among these conceptual possibilities, cell implantation in various forms has been the first strategy to be translated from bench to bedside. The promise of cellular cardiomyogenesis and neovascularization, individually or in tandem, offered altogether novel opportunities for treatment, tailored to the underlying pathobiology.

Within past several years, more than a half-dozen early clinical studies have been published, ranging from case reports to formal trials, deploying a range of differing cell-based therapies with the shared objective of improving cardiac repair [25, 28, 75–80]. Clinical follow-up for as long as a year is now available for some patients. Despite their different strategies and cells, and lack of double-blinded controls, these small initial human trials in general point to a functional improvement; yet key questions remain open. Understanding better just why and how grafting works will be essential, alongside needed empirical trials, to engineer the soundest future for regenerative therapy in human heart disease.

Stem cell mobilization and cardiac repair: Mobilizing stem cells to repair MI damaged hearts—early efforts. Though stem cell mobilization occurs naturally after injury, for unknown reasons, the response following MI is underwhelming, with little functional impact. The goal in the RECOVER trial is to amplify this weak response. Theoretically, one can achieve this goal either by mobilizing available pool of patient's bone marrow stem cells, or expanding them in vitro and re-inject them to the patient. This field is still a matter of debates and controversies as the knowledge on the optimization of the conditions for stem cell therapy is developing and progressing. Clinical and experimental data are difficult to synchronize due to considerable variations in the models used in the studies, as well as the techniques applied. Yet, the experiments with animal models moved clinicians to a series of clinical trials. The pioneers of experimental work and their results are presented in Table 11.2.

Though reperfusion agents and balloon angioplasty help restore cardiac blood flow in the critical hours following myocardial infarction, there is still often residual tissue damage that can prevent full recovery of ventricular function. And for those who miss this 12-h therapeutic window, the pathologic remodeling of the heart is well under way within a week, and well established within a month.

To try and reduce, and ideally reverse, such tissue damage in the first few days or weeks following MI, Cleveland Clinic cardiologists are now beginning to test several stem cell-based strategies. One of these, the RECOVER trial, is a phase I clinical trial that is testing whether injections of a drug, which causes a mobilization of stem cells from the bone marrow into blood, is safe and can lead to improved LV function following major heart attack. The other, which is still in preclinical testing, is looking at this same mobilization drug in combination with injections of myoblasts that have been genetically altered to overproduce the signaling protein that guides stem cells to damaged heart tissue. "Since restoring blood flow is not enough, we need to explore new strategies for achieving full cardiac recovery," says *Ellis GS*, Director of Sones Cardiac Catheterization Laboratory at The Cleveland Clinic, and principal investigator of the RECOVER trial. In this ongoing, double-blinded study, patients who've had a large heart attack (LV ejection fraction less than 40%) within the last 48 h are given subcutaneous injections of placebo or G-CSF, a mobilizing agent that increases the level of circulating stem cells by a factor of 5 within 2 days. The patients are followed for 1 year, with their heart function assessed by echocardiography at 30 days and 12 months. Previous animal studies, as well as a European clinical trial (FIRST LINE) in which G-CSF was administered to 26 patients with MIs, showed that such mobilization can lead to modest improvement in left ventricular function. Intra-myocardial skeletal muscle transplantation has been demonstrated to improve cardiac function in chronic heart failure models by regenerating muscle. Under local anesthesia, a muscle biopsy is carried out to collect skeletal cells for culturing. After about 14 days, the cultured myoblasts can be implanted into the post-MI scar during coronary artery bypass grafting of remote myocardial areas. It is hypothesized that the transplanted autologous myoblasts will aid in repairing the injured area and improving cardiac contractility. However, the safety and effectiveness of this procedure has yet to be established by randomized controlled trials.

Smits et al. reported on the procedural and 6-month results of the first percutaneous and stand-alone study ($n=5$) on myocardial repair with autologous skeletal myoblasts [80]. All cell transplantation procedures were uneventful, and no serious adverse events occurred during follow-up. One patient received an implantable cardioverter-defibrillator after transplantation because of asymptomatic runs of non-sustained ventricular tachycardia. Compared with baseline, the left ventricular ejection fraction (LVEF) increased from 36 ± 11 to $41\pm9\%$ (3 months, $p=0.009$) and $45\pm8\%$ (6 months, $p=0.23$). Regional wall analysis by MRI showed significantly increased wall thickening at the target areas and less wall thickening in remote areas (wall thickening at target areas vs. 3 months follow-up: 0.9 ± 2.3 mm vs. 1.8 ± 2.4 mm, $p=0.008$). The authors concluded that this pilot study was the first to demonstrate the potential and feasibility of percutaneous skeletal myoblast delivery as a stand-alone procedure for myocardial repair in patients with post-infarction heart failure. More data are needed to confirm its safety and explain the phenomenon. Ince et al. stated that transcatheter transplantation of autologous skeletal myoblasts for severe left ventricular dysfunction in post-infarction patients is feasible, safe, and promising. These authors further stated that scrutiny with randomized, double-blinded, multicenter trials appears warranted [81]. This is in agreement with the observation of Siminiak et al. who stated that autologous skeletal myoblast transplantation for the treatment of post-infarction heart failure is feasible, and that further research is needed to validate this method in a clinical practice [82].

Wollert et al. reported that injection of autologous bone marrow stem cells into the coronary arteries improved heart function in patients ($n=60$) who have suffered a myocardial infarction. Patients who had undergone successful percutaneous coronary intervention (PCI) were randomized to receive bone marrow stem-cell transfer, injected into the artery supplying the damaged area of the heart, 5 days after PCI or optimal conventional therapy. After 6 months, improved recovery of LVEF was more evident in patients who received stem-cell transfer therapy than in patients treated with standard post-infarction medical care. Mean global LVEF increased by 7% in the stem-cell transfer group compared with 0.7% in the medical group. The improvement was still evident 6 months after the treatment [78]. The authors suggested that autologous bone-marrow cells can be used to enhance left-ventricular functional recovery in patients after acute myocardial infarction (AMI). However, larger trials are needed to address the effect of bone-marrow cell transfer on clinical endpoints such as the incidence of heart failure and survival. In a pilot study ($n=4$), Obradovic et al. reported that transplantation of bone marrow-derived progenitor cells into the infarcted area (3–5 days after infarct) was safe, and feasible, and might improve myocardial function. Follow-up period for these patients ranged from 30 to 120 days after infarct. This protocol is not based upon mobilization. The reason for that is the statement that stem cell concentration caused by mobilization in poor mobilizers is not high enough to ensure efficient homing and engraftment of stem cells in infarcted area. On the other hand, that non-physiological leukocytosis might cause occlusion of myocardial arteries. These investigators also concluded that

further follow-up will show whether this treatment is effective in preventing negative remodeling of the left ventricle and reveal potential late adverse events (arrhyth-
mogenicity and propensity for re-stenosis) [2]. On the other hand, in another pilot
study ($n = 5$), Kuethe et al. reported that intracoronary transplantation of autologous,
mononuclear bone marrow stem cells did not lead to any significant improvement
in myocardial function and physical performance of patients with chronic ischemic
heart disease at 12-month follow-up [83]. Two German studies [28, 77] using
sophisticated cardiac magnetic resonance imaging and F-18-Fluorodeoxyglucose–
positron emission tomography (FDG–PET) showed significant decreasing of infarction area and increasing of myocardial viability in the infarction zone. In all these
clinical trials, which have together 83 patients, there were no reported major adverse
cardiac events, including death, reinfarction and symptomatic heart failure. However,
MAGIC trial [21], the only trial that had control coronarography in study protocol
after 6 months, showed very high incidence of in stent restenosis (5 of 7 patients) in
the group of patients received cell therapy. But, in this trial G-CSF was used for the
mobilization of bone marrow progenitors to peripheral blood, and this fact might be
the reason for the high incidence of in-stent restenosis [84].

Zohlnhöfer et al. from the Technische Universität, Munich, Germany, conducted
a randomized, double-blind, placebo-controlled study (REVIVAL-2) to assess the
value of G-CSF treatment in a large group of patients following a heart attack. The
114 patients, diagnosed with ST-segment elevation AMI (a certain pattern on an
electrocardiogram indicating a heart attack), had successful reperfusion (restoration
of blood flow) to the heart by PCI (procedures such as angioplasty in which a cath-
eter-guided balloon is used to open a narrowed coronary artery) within 12 h after
onset of symptoms. Patients were randomly assigned to receive by injection either
a daily dose of 10 µg/kg of G-CSF or placebo (an inactive substance) for 5 days.
The patients were treated between February 2004 and February 2005. Treatment
with G-CSF produced a significant mobilization of bone marrow stem cells.
However, the researchers found that this did not alter infarct size (area of damage)
or left ventricular function after a heart attack. "Moreover, in contrast to other studies, no increase in the risk of restenosis (narrowing again of an artery after treatment) or major adverse cardiac events was observed with G-CSF treatment," the
authors write [86]. "The REVIVAL-2 trial had a cohort that was larger than all 3
previous trials taken together and had a relatively long follow-up period based on
sensitive assessment methods of left ventricular function and infarct size. In conclusion, use of G-CSF therapy to mobilize bone marrow-derived stem cell did not
improve left ventricular recovery in patients with AMI after successful mechanical
reperfusion, in this study [87]. In humans, the biological limitations to cardiac
regenerative growth create both a clinical imperative—to offset cell death in acute
ischemic injury and chronic heart failure—and a clinical opportunity; that is, for
using cells, genes, and proteins to rescue cardiac muscle cell number or in other
ways promote more efficacious cardiac repair. Yet, recent experimental studies and
early-phase clinical trials lend credence to the visionary goal of enhancing cardiac
repair as an achievable therapeutic target. The most of the clinical trials and their
outcomes are presented in Tables 11.2 and 15.1.

Table 15.1 Inspiration and driving force for clinical studies: animal models

Donald Orlic (NIH, Bethesda, MD) (mice) (2001) *Nature*	Hamano H (Japan) (2001) (sheep, rabbit model of hindlimb ischemia, Lewis rats-autologous BM transplants)
A. Mouse [27]	B. Sheep
AMI→ same	AMI→ ligation of circumflex arterial branches
Mobilization (noninvasive)	BM biopsy (surgical, invasive method)
Lin– c-kit+ BMC from syngeneic animals	
SCF and G-CSF	
29 BMC in normal to 7,200 in cytokine-treated animals	
3×10^4–2×10^5/5 mL PBS injected 2.5 mL PBS containing cells	4.22×10^8 cells/3 mL injected in ten sites across the infarcted area through the reopened thoracotomy

Basic postulates: The best established source for adult stem cells is the BM. It contains different cell types. HSCs posses plasticity and transdifferentiate into myocardial cells

Cell mobilization: Different aspects. The first hints that cytokine-induced mobilization may be a way to enhance cardiac repair came as an extrapolation of findings of results from efforts to increase EPC levels for neovascularization in another context—hind limb ischemia. Indeed, VEGF [87] and GM-CSF [88] were found to augment EPC levels and improve neovascularization, and subsequent studies documented EPC mobilization by numerous other proangiogenic growth factors—stromal cell-derived factor-1 (SDF-1), angiopoietin-1, placental growth factor, and erythropoietin [89–91]. A wide array of interventions even more accessible clinically than growth factor administration enhance the number of circulating EPCs in adults, including treatment with HMG CoA reductase inhibitors (statins) and estrogens as well as exercise [92–94]. The effect of EPC upon myocardial revascularization is documented in dogs and might be applied in humans, as well. The CXR4 receptor-SDF-1 axis functions in homing HSCs and EPCs to the bone marrow microenvironment, while by employing hemotactic isolation to SDF-1 gradient, these cells can be mobilized.

Most studies confirmed an improvement in endothelial regeneration or neovascularization by mobilizing agents. However, such functional improvements may not rely entirely on EPC mobilization but may also—at least in part—be explained by direct proangiogenic or antiapoptotic effects. Hence, as discussed as a recurring theme in this review, the existence of known (and potential unknown) pleiotropic modes of action complicates the interpretation of regenerative therapies, even in cases where the beneficial effect is clear-cut and assured. A shift in emphasis from the heart's vessels to the heart itself was prompted by the report that bone marrow-derived cells can differentiate into cardiomyocytes when injected into injured myocardium and regenerate the heart effectively [27]. Based on this discovery, hematopoietic stem cell-mobilizing factors—G-CSF and SCF (Kit ligand)—were used to improve cardiac regeneration experimentally (Orlic et al.), which quickly led to the initiation of clinical trials studying the ability of G-CSF to mobilize stem/progenitor cells in patients with coronary artery disease. This cytokine is used routinely in the treatment of humans, e.g., to help in harvesting cells for bone marrow transplantation. Although results from these first small trials do not permit any conclusion of efficacy, the safety of G-CSF in AMI has already come into question [26]. The observed increase in restenosis may be partially explained by the study design (which precluded the standard clinical practice of promptly stenting the obstructed vessel), but the rise in leukocyte number to leukemic levels may be directly responsible, via plaque growth or destabilization. Adverse vascular events have also been attributed to G-CSF in patients with intractable angina who were not candidates for revascularization and even in patients without cardiac disease. In the future, it may be preferable to use strategies that augment circulating progenitor cells (EPC) without causing massive inflammation. A second opened question regarding systemic mobilization is whether enough progenitor cells will home where needed, to the sites of cardiac injury [95]. Systemically administered human progenitor cells were predominantly trapped by the spleen when given to athymic nude rats [96], and cardiac regeneration elicited by the treatment with G-CSF plus SCF was documented only for animals lacking a spleen [26]. The use of leukocyte-mobilizing cytokines might be most worthwhile combined with selective enhancements of progenitor cell homing or as a prelude to isolating cells for local delivery [95].

The potential use of AMD3100 as a mobilization factor for the treatment of AMI. In a mouse model of cardiac infarct caused by ligation of coronary artery, transplanted bone marrow-derived cells were shown to protect against MI by regenerating new myocardial tissue [27]. Subsequent studies in the same model showed that treatment with stem-cell mobilizing cytokines stem cell factor (CSF) and G-CSF resulted in significant tissue regeneration and improvement in cardiac function following cardiac infarct [26]. Though the exact mechanism by which stem cells exert a protective effect resulting in observed tissue repair and improved heart function is a matter of debate and still has to be defined, the evidence is conclusive that mobilized bone marrow-derived adult stem cells are responsible for this tissue regeneration and protection in these disease models. By extrapolation it seems reasonable to assume that AMD3100 (AnorMed) mobilized cells may have similar properties to G-CSF mobilized cells. A number of in vitro pharmacological studies have shown that AMD3100 is a potent and selective antagonist of the CXR4 receptor. The CXR4 receptor-SDF-1 axis functions in homing HSCs and EPCs to the bone marrow microenvironment, while by employing hemotactic isolation to SDF-1 gradient, these cells can be mobilized. Hence, AMD3100 is a potent and selective antagonist of the CXCR4 chemokine receptor that blocks binding of its cognate ligand, stromal cell-derived factor 1 alpha (SDF-1 alpha). It is currently under investigation in the treatment of multiple myeloma (MM), non-Hodgkin lymphoma (NHL), and other hematopoietic malignancies, prior to the high dose chemotherapy. Concurrently, AMD3100 is under investigation for the mobilization of stem cells in the treatment of AMI [97, 98].

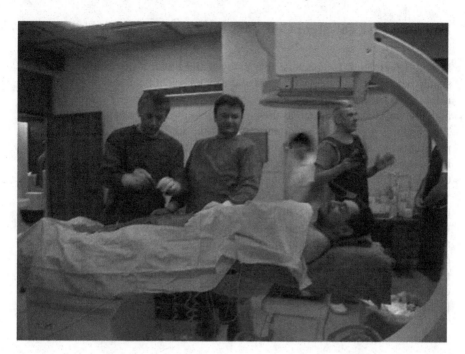

Fig. 15.4 Bone marrow aspiration and filtering of bone marrow after intervention on patient's bedside

Potential risks of mobilization: In dogs, Vulliet et al., have found that injection of BM derived mesenchymal stromal cells caused acute and subacute myocardial microinfarction by occluding coronary circulation [99]. They showed that the size of mesenchymal stromal cells was twofold larger then freshly prepared monoucleated cells. In MAGIC study G-CSF alone did not have any effect on cardiac function.

A mild increase in cardiac enzymes was found after cell infusion by Kang et al. in the same study [21]. The results of Fig. 15.4 published in these two papers suggest that more investigation of potential complications should be done before mesenchymal stromal cells or G-CSF-mobilized cells are routinely injected into the arterial circulation of the patients. In humans, there is the lack of regeneration of myocardium by autologous BM mononuclear cell (MNC) with large anterior myocardial infarctions [82]. Infarct size is a major determinant of morbidity and mortality, as massive infarcts affecting 40% or more of the left ventricle in patients are associated with intractable cardiogenic shock or the rapid development of congestive heart failure. Lack of regeneration due to insufficient number of stem cells gave only 4×10^7 MNCs while in TOPCARE-AMI study (2002) performed by the same author, fourfold higher amount of MNCs with a presumably higher amount of cells with stem cell characteristics was given [77].

Potential Risks of Stem Cell Therapy

It is still premature to conclude that every patient with a myocardial infarction should be treated with stem cells. One of the major scientific merits of these studies is that it has to be investigated—in a rigorously controlled manner—with respect to both: the possible role as well as the limitations of the administration of stem cells. T he findings are thus an important driving force for further targeted clinical and pre-clinical research. Adult human stem cells that are intrinsic to various tissues have been described and characterized, some of them only recently. These cells are capable of maintaining, generating, and replacing terminally differentiated cells within their own specific tissue as a consequence of physiologic cell turnover or tissue damage due to injury. Still, there are the risks of stem-cell therapy in AMI, as follows: The intracoronary application of the BMC carries the risk that cells will not attach to the vasculature and migrate into the myocardium but be washed out after deflation of the obstructing balloon if the balloon is used. Labeling of BMCs prior to injection should be performed allowing the demonstration of cell attachment. The effect of clinical outcome followed through PET from the study of Obradovic et al. [100] is shown in Fig. 15.5.

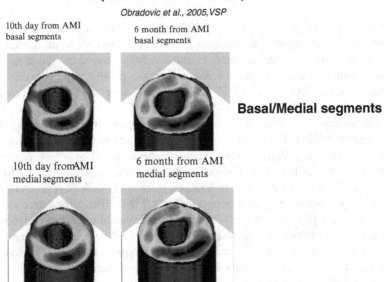

VISIBLE RESULTS OF MNC THERAPY

(STEM CELL IMPLANT)

Obradovic et al., 2005, VSP

10th day from AMI
basal segments

6 month from AMI
basal segments

Basal/Medial segments

10th day from AMI
medial segments

6 month from AMI
medial segments

Fig. 15.5 SPECT of the first patient 10 days and 6 months after AMI

Summary and Conclusions on the Stem Cell Therapy in AMI

Bone marrow is, at present, the most frequent source of cells used for clinical cardiac repair. It contains a complex assortment of stem and progenitor cells, including HSCs; the so-called side population (SP) cells, defined by their ability to expel a Hoechst dye, which account for most if not all long-term self-renewal and reconstitute the full panoply of hematopoietic lineages after single-cell grafting; a subset of MSCs or stromal cells, which are already defined (MSCs); multipotential adult non-hematopoietic progenitor cells (MAPCs, MIAMI, e.g. VSELCs), which can differentiate into all possible lineages, a fraction of BM EPC cells shown to induce revascularization, and a fraction of already defined non-hematopoietic tissue committed cells (TCSC) discovered by Ratajczak and colleagues [49]. These, TCSCs, circulate at the highest level and thus accumulate in BM during rapid body growth, and become a reserve pool of stem cells for tissue/organ regeneration [49]. They are chemoattracted from peripheral blood to damaged organs by SDF-1,that becomes highly expressed in injured tissues [52]. For therapeutical purposes, bone marrow is aspirated under local or general anesthesia, the entire MNC fraction is obtained (a heterogeneous mix of the above-mentioned cells), or specific subpopulations are purified, and isolated cells are injected into the heart without need of further ex vivo expansion. Expansion in cell culture could be desirable or essential, though, if defined but minute subpopulations prove to be advantageous. Different possibilities are on the horizon for expansion. As VSELSC are probably present in humans, cloning will be one of the options, (Youn-sup Yan et al.) since it is difficult to propagate them in culture due to the short self-renewal time caused by accelerated differentiation induced by culture media [67]. The humanized adaptive mouse model system is also proposed as one of the possibilities, but it requires further research to become fully acceptable in clinical arena [101]. So, it seems that what we are using right now for the "stem cell therapy" in MI and other diseases treated in this way, are MNC and progenitor cells. Only a small fraction within them belongs to multipotent non-hematopoietic cells. Therefore, there is no doubt that this kind of therapeutical approach will need a fine tuning in the future for each particular purpose.

There are a lot of opened questions considering progenitor cell therapy in the treatment of ischemic heart disease. What is the exact mechanism and extent of benefit of bone marrow-derived progenitor cells infusion into the infarct related artery? Do we need some special cell compartment, and some selection of bone marrow cells before the application into the myocardium? Do we need cytokine preparation of bone marrow and myocardium to accelerate the number of needed progenitors and to improve homing and engraftment of these cells into the ischemic myocardium? Do we need a strictly determined number of EPCs for better revascularization as it is suggested by animal studies? The first clinical trials showed encouraging results of this therapy, the possibility to enhance regeneration of damaged myocardial tissue, which seemed to be impossible just a few years ago. Yet, the precise mechanism by which stem cells exert a protective effect resulting in the observed tissue repair and improvement in heart function is also a matter of debate.

A number of possible mechanisms have been proposed: (a) transdifferentiation of stem cells into cells of other lineages such as endothelial cells or cardiomyocytes resulting in the formation of new tissue; (b) mobilization of tissue-specific stem/progenitor cells from the BM that home to the damaged tissue and participate in tissue/organ regeneration [102]; (c) fusion of the stem cells with cells of the target tissue giving rise to new cells [103, 104], and/or (d) creation of a millue, perhaps by secretion of growth factors, that enhances regeneration of endogenous cells [105, 106]. Though the exact mechanism still has to be defined the evidence is conclusive that mobilized bone marrow-derived adult stem cells are responsible for the tissue regeneration and protection in these disease models.

In summary, current challenges for cell-based therapy in cardiac repair include identifying the origins of the novel cardiac progenitor and stem cells found within the heart, pinpointing the biologically active cells from bone marrow and other mixed populations, optimizing cell mobilization and homing, augmenting grafted cells' survival, defining the cues for cardiac differentiation, promoting donor cell proliferation ex vivo (or, if safe, in vivo), and exploiting cell therapy as a platform for secretory signals. In conclusion: we need a lot of basic research and randomized clinical trials to define the exact role of this probably revolutionary therapy for ischemic heart disease.

References

1. Kocher AA, Schuster MD, Szabolcs MJ, Takuma S, Burkoff D, Wang J et al (2001) Neovascularization of ischemic myocardium by human bone-marrow-derived angioblasts prevents cardiomyocyte apoptosis, reduced remodelling and improves cardiac function. Nat Med 7:430–436
2. Obradovic S, Balint B, Rusovic S, Ristic-Angelkov A, Romanovic R, Baskot B et al (2004) The first experience with autologous bone marrow derived progenitor cell transfer for myocardial regeneration after acute infarction. Anesth Reanim Transf 32(1–2):39–50
3. Dai W, Kloner RA (2006) Myocardial regeneration by embryonic SC transplantation: present and future trends. Expert Rev Cardiovasc Ther 4(3):375–383
4. Rogers I, Casper RF (2004) Umbilical cord blood SCs. Best Pract Res Clin Obstet Gynaecol 18(6):893–908
5. Rangappa S, Fen C, Lee EH, Bongso A, Sim EK (2003) Transformation of adult mesenchymal stem cells isolated from the fatty tissue into cardiomyocytes. Ann Thorac Surg 75:775–779
6. Lakshmipathy U, Pelacho B, Sudo K, Linehan JL, Coucouvanis E, Kaufman DS, Verfaillie CM (2004) Efficient transfection of embryonic and adult stem cells. Stem Cells 22(4):531–543
7. Hierlihy AM, Seale P, Lobe CG, Rudnicki MA, Megeney LA (2002) The post-natal heart contains a myocardial stem cell population. FEBS Lett 520:239–243
8. Yau TM, Tomita S, Weisel RD, Jia Z-Q, Tumiati LC, Mickle DAG et al (2003) Beneficial effect of autologous cell transplantation on infarcted heart function: comparison between bone marrow stromal cells and heart cells. Ann Thorac Surg 75:169–177
9. Dowell JD, Rubart M, Pasumarthi KBS, Soonpaa MH, Field LJ (2003) Myocyte and myogenic stem cell transplantation in the heart. Cardiovasc Res 58:333–347
10. Shintani S, Murohara T, Ikeda H, Uenoi T, Honma T, Katoh A et al (2001) Mobilization of endothelial progenitor cells in patients with acute myocardial infarction. Circulation 103:2776–2779

11. Kuznetsov SA, Mankani MH, Gronthos S, Satomura K, Bianco P, Robey PG (2001) Circulating skeletal stem cells. J Cell Biol 153:1133–1139
12. Brehm M, Zeus T, Strauer BE (2002) Stem cells-clinical application and perspectives. Herz 27(7):611–620
13. Penn MS, Francis GS, Ellis SG, Young JB, McCarthy PM, Topol EJ (2002) Skeletal myoblast transplantation for the treatment of damage myocardium. Prog Cardiovasc Dis 45:21–32
14. Rafii S, Lyden D (2003) Therapeutic stem and progenitor cell transplantation for organ vascularization and regeneration. Nat Med 9(6):702–712
15. Khakoo AY, Finkel T (2005) Endothelial progenitor cells. Annu Rev Med 56:79–101
16. Hassink RJ, de la Rivere AB, Mummery CL, Doevendans PA (2003) Transplantation of cells for cardiac repair. J Am Coll Cardiol 41(5):771–777
17. Makino S, Fukuda K, Miyoshi S, Konishi F, Kodama H, Pan J et al (1999) Cardiomyocytes can be generated from stromal cells in vitro. J Clin Invest 103:697–705
18. Murry CE, Wiseman RW, Schwartz SM, Hauschka SD (1996) Skeletal myoblasts transplantation for repair of myocardial necrosis. J Clin Invest 98:2512–2523
19. Yeh ETH, Zhang S, Wu HD, Körbling M, Willerson JT, Estrov Z (2003) Transdifferentiation of human peripheral blood CD34+-enriched cell population into cardiomyocytes, endothelial cells, and smooth muscle cells in vivo. Circulation 108:2070–2073
20. Obradović S, Rusović S, Dinčić D, Gligić B, Baškot B, Balint B, i sar (2003) Autologe pluripotentne progenitorne ćelije u lečenju ishemijske bolesti srca. Vojnosanit Pregl 60(6):725–31
21. Kang H-J, Kim H-S, Zhang S-Y, Park K-W, Cho H-J, Koo B-K et al (2004) Effects of intracoronary infusion of peripheral blood stem-cells mobilized with granulocyte-colony stimulating factor on left ventricular systolic function and restenosis after coronary stenting in myocardial infarction: the MAGIC cell randomized clinical trial. Lancet 363:751–756
22. Avilés FF. Conventional postreperfusion therapy versus intracoronary bone marrow stem cell transplantation or mobilization after STEMI. The TECAM randomised trial. First international symposium on cell therapy for cardiac diseases, Valladolid, Spain, March 26 2004
23. Tse HF, Kwong YM, Chan JKF, Lo G, Ho C-L, Lau C-P (2003) Angiogenesis in ischaemic myocardium by intramyocardial autologous bone marrow mononuclear cell implantation. Lancet 361:47–49
24. Fuchs S, Satler L, Kornowski R, Okubagzi P, Weisz G, Baffour R et al (2003) Catheter-based autologous bone marrow myocardial injection in no-option patients with advanced coronary artery disease. J Am Coll Cardiol 41:1721–1724
25. Perin EC, Dohmann HFR, Borojevic R, Silva SA, Sousa ALS, Mesquita CT et al (2003) Transendocardial, autologous bone marrow cell transplantation for severe, chronic ischemic heart failure. Circulation 107:2294–2302
26. Orlic D, Kajstura J, Chimenti S, Limana F, Jakoniuk I, Quani F et al (2001) Mobilized bone marrow cells repair the infarcted heart, improving function and survival. Proc Natl Acad Sci U S A 98:10344–10349
27. Orlic D, Kajstura J, Chimenti S, Jakoniuk I, Anderson SM, Li B et al (2001) Bone marrow cells regenerate infarcted myocardium. Nature 410:701–705
28. Strauer BE, Brehm M, Zeus T, Köstering M, Hernandez A, Sorg RV et al (2002) Repair of infarcted myocardium by autologous intracoronary mononuclear bone marrow cell transplantation in humans. Circulation 106:1913–1918
29. Aicher A et al (2003) Assessment of the tissue distribution of transplanted human endothelial progenitor cells by radioactive labeling. Circulation 107:2134–2139
30. Beauchamp JR, Morgan JE, Pagel CN, Partridge TA (1999) Dynamics of myoblast transplantation reveal a discrete minority of precursors with stem cell-like properties as the myogenic source. J Cell Biol 144:1113–1122
31. Liu F, Pan X, Chen G, Jiang D, Cong X, Fei R, Wei L (2006) Hematopoietic SCs mobilized by granulocyte colony-stimulating factor partly contribute to liver graft regeneration after partial orthotopic liver transplantation. Liver Transpl 12(7):1129–1137

32. Perin EC et al (2003) Transendocardial, autologous bone marrow cell transplantation for severe, chronic ischemic heart failure. Circulation 107:2294–2302
33. Taylor DA et al (1998) Regenerating functional myocardium: improved performance after skeletal myoblast transplantation. Nat Med 4:929–933
34. Hutcheson KA et al (2000) Comparison of benefits on myocardial performance of cellular cardiomyoplasty with skeletal myoblasts and fibroblasts. Cell Transplant 9:359–368
35. Thompson RB et al (2003) Comparison of intracardiac cell transplantation: autologous skeletal myoblasts versus bone marrow cells. Circulation 108(suppl 1):II264–II271
36. Balsam LB et al (2004) Haematopoietic stem cells adopt mature haematopoietic fates in ischaemic myocardium. Nature 428:668–673
37. Murry CE et al (2004) Haematopoietic stem cells do not transdifferentiate into cardiac myocytes in myocardial infarcts. Nature 428:664–668
38. Nygren JM et al (2004) Bone marrow-derived hematopoietic cells generate cardiomyocytes at a low frequency through cell fusion, but not transdifferentiation. Nat Med 10:494–501
39. Jin DK, Shido K, Kopp HG, Petit I, Shmelkov SV, Young LM et al (2006) Cytokine-mediated deployment of SDF-1 induces revascularization through recruitment of CXCR4+ hemangiocytes. Nat Med 12(5):557–67. Erratum in: Nat Med 12(8):978
40. Rehman J, Li J, Orschell CM, March KL (2003) Peripheral blood endothelial progenitor cells are derived from monocyte/macrophages and secrete angiogenic growth factors. Circulation 107:1164–1169
41. Kinnaird T, Stabile E, Burnett MS, Lee CW, Barr S, Fuchs S et al (2004) Marrow-derived stromal cells express genes encoding a broad spectrum of arteriogenic cytokines and promote in vitro and in vivo arteriogenesis through paracrine mechanism. Circ Res 94:678–685
42. Zhang S, Zhang P, Guo J, Jia Z, Ma K, Liu Y et al (2004) Enhanced cytoprotection and angiogenesis by bone marrow cell transplantation may contribute to improved ischemic myocardial function. Eur J Cardiothorac Surg 25:188–195
43. Hirata K, Li TS, Nishida M, Ito H, Matsuzaki M, Kasaoka S, Hamano K (2002) Autologous bone marrow cell implantation as therapeutic angiogenesis for ischeminc hindlimb in diabetic rat model. Am J Physiol Heart Circ Physiol 28(1):H66–H70
44. Adams BG, Scadden DT (2006) The hematopoietic stem cells in its place. Nat Immunol 7:333–337
45. Adams BG, Chabner KT, Alley JR, Scadden D et al (2006) Stem cell engraftment at the endosteal niche is specified by the calcium-sesning receptor. Nat Lett 439:599–603. doi:10.1038/nature0424
46. Wojakowski W, Tendera M, Michalowska A, Majka M, Kucia M, Ratajczak MZ et al (2004) Mobilization of CD34/CXCR4+, CD34/CD117+, c-met+ stem cells, and mononuclear cells expressing early cardiac, muscle, and endothelial markers into peripheral blood in patients with acute myocardial infarction. Circulation (Genetics) 110:3213–3220
47. Kucia M, Dawn B, Hunt G, Guo Y, Wysoczynski M, Majka M et al (2004) Cells expressing early cardiac markers reside in the bone marrow and are mobilized into the peripheral blood after myocardial infarction. Circulation Res 95(12):1191–1199
48. Kucia M, Reca R, Jala V, Dawn B, Ratajczak J, Ratajczak MZ (2005) Bone marrow as home of heterogeneous populations of nonhematopoietic stem cells. Leukemia 19:1118–1127
49. Kucia M, Ratajczak J, Ratajczak ZM (2005) Bone marrow as a source of circulating CXR4+ tissue-commited stem cells. Biol Cell 97:133–146
50. Kucia M, Ratajczak J, Ratjczak MX (2005) Are bone marrow cells plastic or heterogeneous-that is the question. Exp Hematol 33(6):613–623
51. Ratajczak J, Miekus K, Kucia M, Zhang J, Reca R, Dvorak P, Ratajczak MZ (2006) Embryonic stemcell-derived microvesicles reprogram hematopoietic progenitors: evidence for horizontal transfer of mRNA and protein delivery. Leukemia 20:847–856
52. Kucia M, Wojakowski W, Ryan R, Machalinski B, Gozdzik J, Majka M, Baran J, Ratajczak J, Ratjczak MZ (2006) The migration of bone marrow-derived non-hematopoietic tissue-commited stem cells is regulated in and SDF-1-, HGF-, and LIF dependent manner. Arch Immunol Ther Exp 54(2):121–135

53. Kucia M, Zhang YP, Reac R, Wysoczynski M, Machalinski B, Majka M, Ildstad ST, Ratajczak JU, Chields CB, Ratajczak MZ (2006) Cells enriched in markers of neural tissue-commited stem cells reside in the bone marrow and are mobilized into the peripheral blood following stroke. Leukemia 20:18–28

54. Kucia M, Reca R, Campbell FR, Surma-Zuba E, Majka M, Ratajczak M, Ratajczak MZ (2006) A population of very small embryonic-like (VSEL) CXR4+ SSEA-1+ Oct4+ stem cells identified in adult bone marrow. Leukemia 20:857–869

55. Wagers AJ, Sherwood RI, Christensen JL, Weissman IL (2002) Little evidence for developmental plasticity of adult hematopoietic stem cells. Science 297:2256–2259

56. Murry CE, Soonpaa MH, Reinecke H, Nakajima H, Rubart M, Pasumarthi KBS et al (2004) Haematopoietic stem cells do not transdifferentiate into cardiac myocytes in myocardial infarction. Nature 428:664–668

57. Ziegelhoeffer T, Fernandez B, Kostin S, Heil M, Voswinckel R, Helisch A et al (2004) Bone marrow-derived cells do not incorporate into adult growing vasculature. Circ Res 94:230–238

58. Menasche P (2004) Cellular transplantation: hurdles remaining before widespread clinical use. Curr Opin Cardiol 19:154–161

59. Levy JS, Stroomza M, Melemed E, Offen D (2004) Embryonic and adult stem cells as a soutce for cell therapy in Parkinson's disease. J Mol Neurosci 24(3):353–386

60. De Copi P, Bartsch G Jr, Minhaj MS, Atala A et al (2007) Isolation of amniotic stem cell lines with potential for therapy. Nat Biotechnol 25:100–106

61. Kim J, Lee Y, Hwanf KJ, Kwon HC, Kim SK, Cho DJ, Kang SG, You J (2007) Human amniotic fluid-derived stem cells have characteristics of multipotent stem cells. Cell Prolif 40(1):75–90

62. Jendelova P, Herynek V, Urdzikova L, Glogarova K, Kroupova J, Anderson B, Vryja V, Burtain M, Hajek M, Sykova J (2004) Magnetic resonance tracking of transplanted bone marrow and embryonic stem cells labeled by iron oxide nanoparticles in rat brain and spinal cord. J Neurosci Res 76:232–243

63. Marshall CT, Lu C, Winstead W, Zhanf X, Xiao M, Harding G, Klueber KM, Roisen FJ (2006) The therapeutic potential of human olfactory-derived stem cells. Histol Histopathol 21:633–643

64. Lindwall O, Kokaia Z, Martinez-Serrano A (2006) SCs for the treatment of neurological disorders. Nature 441:1094–1096

65. D'Ippolito G, Sylma D, Howadr GA, Philippe M, Roos BA, Schiller PC (2004) Marrow-isolated adult multilineage inducible (MIAMI) cells, a unique population of postnatal young and old human cells with extensive expansion and differentiation potential. J Cell Sci 117:2971–2981

66. Serafini M, Dylla SJ, Oki M, Heremans Y, Tolar J, Jiang Y et al (2007) Hematopoietic reconstitution by multipotent adult progenitor cells: precursors to long-term hematopoietic stem cells. J Exp Med 204(1):129–139

67. Young-sup Y, Wecker A, Heyd L, Park JS, Douglas W, Losordo DW et al (2005) Clonally expanded novel multipotent stem cells from human bone marrow regenerate myocardium after myocardial infarction. J Clin Invest 115(2):326–338

68. Balint B (2004) SCs—unselected or selected, unfrozen or cryopreserved: marrow repopulation capacity and plasticity potential in experimental and clinical settings. Maked Med Pregl 58(suppl 63):22–24

69. Pelacho B, Aranguren XL, Mazo M, Abizanda G, Gavira JJ, Clavel C et al (2007) Prosper plasticity and cardiovascular applications of multipotent adult progenitor cells. Nat Clin Pract Cardiovasc Med 4(suppl 1):S15–S20

70. Müller P, Pfeiffer P, Koglin J, Schafers HJ, Seeland U, Janzen I et al (2002) Cardiomyocytes of noncardiac origin in myocardial biopsies of human transplanted hearts. Circulation 106:31–35

71. Poss KD, Wilson LG, Keating MT (2002) Heart regeneration in zebrafish. Science 298:2188–2190

72. Bettencourt-Dias M, Mittnacht S, Brockes JP (2003) Heterogeneous proliferative potential in regenerative adult newt cardiomyocytes. J Cell Sci 116:4001–4009

73. Pasumarthi KB, Nakajima H, Nakajima HO, Soonpaa MH, Field LJ (2005) Targeted expression of cyclin D2 results in cardiomyocyte DNA synthesis and infarct regression in transgenic mice. Circ Res 96:110–118

74. Oh H, Wang SC, Prahash A et al (2003) Telomere attrition and chk2 activation in human heart failure. Proc Natl Acad Sci U S A 100:5378–5383

75. Olson EN, Schneider MD (2003) Sizing up the heart: development redux in disease. Genes Dev 17:1937–1956

76. Menache P, Hagege AA, Vilquin JT, Desnos M, Abergel E, Pouzet B et al (2003) Autologous skeletal myoblast transplantation for severe postinfarction left ventricular dysfunction. J Am Coll Cardiol 41:1078–1083

77. Assmus B, Schachinger V, Teupe C, Britten M, Lehmann R, Döbert N et al (2002) Transplantation of progenitor cells and regeneration enhancement in acute myocardial infarction (TOPCARE-AMI). Circulation 106:3009–3017

78. Stamm C, Westphal B, Kleine HD, Petzsch M, Kittner C, Klinge H et al (2003) Autologous bone-marrow stem-cell transplantation for myocardial regeneration. Lancet 361:45–46

79. Wollert KC, Meyer GP, Ringes-Lichtenberg S, Drexler H et al (2004) Intracoronary autologous bone-marrow cell transfer after myocardial infarction: the BOOST randomized controlled clinical trial. Lancet 364(9429):141–148

80. Smits PC et al (2003) Catheter-based intramyocardial injection of autologous skeletal myoblasts as a primary treatment of ischemic heart failure: clinical experience with six-month follow-up. J Am Coll Cardiol 42:2063–2069

81. Ince H, Petzsch M, Dieter Keline H et al (2005) Preservation from left ventricular remodeling by front-integrated revascularization and stem cell liberation in evolving acute myocardial infarction by use of granulocyte-colony-stimulating factor (FIRSTLINE-AMI). Circulation 112:3097–3106

82. Siminiak T, Kalawski R, Fiszer D, Jerzykowska O, Rzezniczak J, Rozwadowska N et al (2004) Autologous skeletal myoblast transplantation for the treatment of postinfarction myocardial injury: phase I clinical study with 12 months of follow-up. Am Heart J 148(3):531–537

83. Kuethe F, Figulla HR, Herzau M, Voth M, Fritzenwanger M, Opfermann T, Pachmann K, Krack A, Sayer HG, Gottschild D, Werner GS (2005) Treatment with granulocyte colony-stimulating factor for mobilization of bone marrow cells in patients with acute myocardial infarction. Am Heart J 150(1):115

84. Welt FGP, Edelman ER, Simon DI, Rogers C (2000) Neutrophil, not macrophage, infiltration precedes neointimal thickening in balloon-injured arteries. Arterioscler Thromb Vasc Biol 20:2553–2558

85. Zohlnhofer D, Kastrati A, Schomig A (2007) Stem cell mobilization by granulocyte-colony-stimulating factor in acute myocardial infarction: lessons from the REVIVAL-2 trial. Nat Clin Pract Cardiovasc Med 4(suppl 1):S106–S109

86. Zohlnhofer D, Kastrati A, Schomig A et al (2006) Stem cell mobilization by granulocyte colony-stimulating factor in patients with acute myocardial infarction: a randomized controlled trial. JAMA 295(9):1003–1010

87. Asahara T, Takahashi T, Masuda H et al (1999) VEGF contributes to postnatal neovascularization by mobilizing bone marrow-derived endothelial progenitor cells. EMBO J 18:3964–3972

88. Takahashi T et al (1999) Ischemia and cytokine-induced mobilization of bone marrow-derived endothelial progenitor cells for neovascularization. Nat Med 5:434–438

89. Hattori K, Heisssig B, Tashiro K et al (2001) Plasma elevation of stromal cell-derived factor-1 induces mobilization of mature and immature hematopoietic progenitor and stem cells. Blood 97:3354–3360

90. Hattori K, Heissig B, Wu Y et al (2002) Placental growth factor reconstitutes hematopoiesis by recruiting vegfr1(+) stem cells from bone-marrow microenvironment. Nat Med 8:841–849

91. Heeschen C, Aicher A, Lehmann R et al (2003) Erythropoietin is a potent physiologic stimulus for endothelial progenitor cell mobilization. Blood 102:1340–1346
92. Dimmeler S, Aicher A, Vasa M et al (2001) HMG-CoA reductase inhibitors (statins) increase endothelial progenitor cells via the PI 3-kinase/Akt pathway. J Clin Invest 108:391–397
93. Laufs U, Werner N, Link A et al (2004) Physical training increases endothelial progenitor cells, inhibits neointima formation, and enhances angiogenesis. Circulation 109:220–226
94. Iwakura A, Leudemann V, Shastry S et al (2003) Estrogen-mediated, endothelial nitric oxide synthase-dependent mobilization of bone marrow-derived endothelial progenitor cells contributes to reendothelialization after arterial injury. Circulation 108:3115–3121
95. Askari AT, Unzek S, Popovich ZB et al (2003) Effect of stromal-cell-derived factor 1 on stem-cell homing and tissue regeneration in ischaemic cardiomyopathy. Lancet 362:697–703
96. Aicher A, Brenner W, Zuhayra M et al (2003) Assessment of the tissue distribution of transplanted human endothelial progenitor cells by radioactive labeling. Circulation 107: 2134–2139
97. Liles WC, Roger E, Broxmeyer HE, Dehner C, Badel K, Calandra G et al (2005) Augmented mobilization and collection of CD34+ hematopoietic cells from normal human volunteers with granulocyte-stimulating factor by single-dose administration ofr AMD3100, a CXXR4 antagonist. Transfusion 45(3):295–300
98. Larochelle A, Krouse A, Metzger M, Orlic D et al (2006) AMD3100 mobilizes hematopoietic stem cells with long-term repopulating capacity in non-numan primates. Blood 107(9):3772–3778
99. Vulliet PR, Greegly M, Halloran SM, MacDonald KA, Kittleson MD (2004) Inracoronary arterial injection of mesenchymal stromal cells and microinfarction in dogs. Lancet 363(9411):783–784
100. Obradovic S, Rusovic S, Dincic D, Gligic B, Baskot B, Balint B et al (2003) [Autologous pluripotent progenitor cells in the treatment of ischemic heart disease]. Vojnosanit Pregl 60(6):725–731
101. Traggiai E, Chicha L, Mazzucchelli L, Bronz L, Piffaretti JC, Lanzavecchia A, Manz MG (2004) Development of a human adaptive immune system in cord blood cell-transplanted mice. Science 304(5667):104–107
102. Ratajzcak MZ, Kucia M, Reca R, Majka M et al (2004) Stem cell plasticity revised: CXR4 positive cells expressing mRNA for early muscle, liver and neural cells "hide out" in the bone marrow. Leukemia 19(1):29–40
103. Vassilopoulos G, Wang PR, Russel DV (2003) Transplanted bone marrow regenerates liver by cell fusion. Nature 422(6934):901–904
104. Wang X, Willenbring H, Akkari Y et al (2003) Cell fusion is the principal source of bone-marrow derived hepatocytes. Nature 422(6934):897–901
105. Chen J, Zhang ZG, Li Y, Wang L et al (2003) Intravenous administration of human bone marrow stromal cells induces angiogenesis in the ischemic boundary zone after stroke in rats. Circ Res 92(6):692–699
106. Prockop DJ, Gregory CA, Spees JL (2003) One strategy for cell and gene therapy; harnessing the power of adult stem cells to repair tissue. Proc Natl Acad Sci U S A 200(suppl 1): 11917–11923

Chapter 16
Stem Cells in Neurodegenerative Diseases. Part I: General Consideration (The Old Idea or A New Therapeutic Concept?)

*Mankind is searching for a key to longevity
and there is no doubt that stem cells could be
an important answer to this problem.*

– Ratajczak Marius

The pluripotency of stem cells from different sources is the subject of controversies and criticism, but as one of the most prominent features of these cells is also envisioned as a powerful therapeutic approach. Adult stem cells reside in most mammalian tissues, but the extent to which they contribute to normal homeostasis and repair varies widely.

Types of Stem Cells and Spectrum of Their Application in Clinical Arena

Types of Stem Cells According to Their Functionality and Their Importance for Neural Tissue

According to this criterion, stem cells can be divided into two distinct categories: normal and cancer stem cells.

1. *Normal stem cells* are immature cells that can replicate, or renew themselves, can divide indefinitely to produce more copies of themselves, and are able to differentiate or mature into all the cells that an organism or a particular organ system needs. Each stem cell is unspecialized, but it can produce progeny that mature into the various cell types of, say, the brain or the immune system. Once this maturation occurs, the stem cell heirs may divide rapidly but only a limited number of times (1–3).

M. Pavlovic and B. Balint, *Stem Cells and Tissue Engineering*,
SpringerBriefs in Electrical and Computer Engineering,
DOI 10.1007/978-1-4614-5505-9_16, © The Author(s) 2013

2. *Cancer stem cells*. Finding cancers' stem cells is a rapidly growing area of research. These cancer-causing cells, which make up a tiny fraction of cells within tumors, according to research data have properties similar to those of normal stem cells and are considered arising from them (3). However, cancer stem cells make up only a tiny number of the total cancer cells in a leukemia patient, which makes the cells next to impossible to find (4). Therefore, it seems that promise of this line of research can only be realized by studying adult stem cells as well as embryonic stem cells (ESCs). The latter are, as we know, still ethical problem and therefore substantially controversial because an early embryo is destroyed when researchers remove stem cells from it. While in many studies volunteers could provide samples, that will not be the case for all types of diseases; an alternative is to take the stem cells from embryos that carry a genetic defect for specific diseases and grow them in a larger number as it was done with patients with chronic myelogenous leukemia (CML) by Weinberg et al. (4). Researchers have traditionally thought of cancer as a collection of cells, all growing exponentially. Weinberg et al. demonstrated convincingly in their study that the model is wrong, since only a few cells were endowed with the ability to replicate (4). It has profound implications for how we think tumors evolve and how we treat tumors. So, conventional cancer therapies do an effective job, killing the majority of cells within the tumor, but they may miss cancer stem cells, according to this new research. As a result, cancers often recur (4–7). The reason is (among others) in the fact that clinicians are reinjecting also cancer cells with healthy stem cells during reinfusion after aphaeresis collection. They accumulate and renew with a time to the critical level causing relapse or death.

Types of Stem Cells According to the Available Sources for the Application in Clinical Arena

(1) *Embryonic stem cells*, (2) *Fetal stem cells*, (3) *Cord blood stem cells, and* (4) *Adult stem cells of bone marrow and different tissues and organs* within an adult organism

Therapeutic Application of Each Particular Type of Stem Cells: Advantages and Limitations

1. *Embryonic stem cells: from therapeutic cloning to stembrids, and back, around the world*
   In 2004 and 2005, South Korean team announced the use of therapeutic cloning to create human ESCs that were genetically and therefore immunologically matched to specific people (the critical issue that we really wish to overcome in

stem cell therapy). The Somatonuclear transfer as a part of therapeutic cloning is a technique in which the patient's nucleus content (DNA) is injected into an enucleated unfertilized egg and used to generate ESCs which are then cultured and allowed to differentiate, following transplantation into the patient. The entire procedure is called therapeutic cloning, since it does supply the source of ESC with patient's HLA system (clones identical immunological features in the organism as the patient already possesses and therefore does not require immunosuppression). The use of such cells may bypass the ethical objections and immunological issues of using ESC and is the future of stem-cell clinical application. The leading scientist Woo Suk Hwang (of Seoul National University) claimed to have created an ESC line from a cloned human embryo for the first time. But there were doubts about the results. The technique requires an adult cell and an egg, and because in Hwang's experiment both came from the same person, it was difficult to prove that embryo really was cloned (8). The process was also very inefficient, taking 242 eggs to create just one ESC line. Then, Hwang has created 11 more ESC lines from cloned embryos in an impressive study that answered all the criticisms of his original study. He has also greatly increased the efficiency of the process: the 11 lines came from just 185 fresh eggs donated by 18 unpaid volunteers, meaning an average of only 17 eggs were needed per ESC line (9). The donor "adult" cells came from patients aged 2–56, with a variety of conditions ranging from spinal injuries to an inherited immune condition. The work proved that matching ESCs can be derived via nuclear transfer from donors of any age and sex. This was an enormous stride in the long journey to determine whether nuclear-transfer-derived human ESC might be eventually suitable for transplantation medicine. However, the chain of events has happened very soon causing Korean group to withdraw their results. Therapeutic cloning is much less controversial in South Korea than in the West. More than 70% of South Koreans agree with therapeutic cloning, whereas a recent poll in the USA suggests that 75% are opposed to it. Furthermore, South Korean law allows the use of fresh eggs from young women who are prepared to donate their eggs by undergoing ovarian stimulation, which can be a risky and painful procedure. In the western countries by contrast and for example, scientists are only allowed to use eggs rejected or left over from in vitro fertilization (IVF) treatment. Apparently, the supply of the eggs is lacking in the West.

While ESCs show great promise for treating many diseases, such as heart disease, diabetes, and Parkinson's, non-matching ESCs would be rejected by patients' immune systems unless they take immunosuppressant drugs. This is why many stem cell researchers are trying to create ESCs that are identical to people's (recipient's) own cells. However, Yuri Verlinsky of the Reproductive Genetics private Institute, in Chicago, claims to have produced patient-matched ESC without resorting to therapeutic cloning. Unlike therapeutic cloning, it uses existing ESCs instead of human eggs, the step much cheaper and easier. Furthermore, because no embryos are destroyed, it would bypass many ethical issues. In fact, if Verlinsky's claims stand up, there might be a much easier way to create matching ESCs. Verlinsky has created 13 ESC lines using his new "stembrid" technique. He has applied for a patent on the so-called *stembrid method (US 2004/0259249)* insisting that he does not intend to stop other researchers from using the method (10). However, Verlinsky did not reveal the details of his technique, which were later on patented. It is known that other teams are trying to attach cells to a surface and spin them until the denser nucleus is forced out (11). The cells from adults are taken afterwards and fused with the enucleated ESCs (11). The idea is that the cytoplasm of an enucleated ESC will reprogram the donor nuclei, turning the fused cell into an ESC, too. The result, as Verlinsky told at a Conference in an oral presentation on preimplantation genetics in London, is new lines of ESCs that are genetically identical to the adult donor (11). He called the new cells *stembrids*. How one could be sure that the stembrids contain only the genetic material of the adult donors? Verlinsky's response was that he fused male adult cells with a female ESC line (not the egg), with resulting male cells. However, his team has not yet tested the HLA proteins on the cell surface to prove that the stembrids are an immune match for the adult donors. By contrast, Hwang's team has shown that the ESCs derived via therapeutic cloning have the surface markers that make them compatible with the donor's immune system. The biggest question, though, is whether the stembrids really are true ESCs. Verlinsky stated (11) that cells express a number of ESC markers, but we do not know which of them. His team has also shown that cells can differentiate into a number of cell types, including heart muscle cells, neurons, and blood stem cells, although the results have not yet been published. Apparently, far more detailed studies will be needed in the future to prove this method, especially since it is not published, yet. Besides looking at further cell surface markers, Verlinsky's team was to show that, like true ESCs, the stembrids are capable of forming cancers called teratomas, which contain a mixture of different cell types, when injected into immune-compromised mice (12). Until these studies are done, we must remain very skeptical. Potentially, these resulting stem cells could be stockpiled for use in researching any number of genetic diseases, including stem cells that will specifically develop seven diseases (including two forms of muscular dystrophy, thalassemia, Fanconi anemia, fragile X syndrome, Marfan syndrome, and a type of neurofibromatosis). Researchers could choose cells directly affected by a disease (parts of the brain impaired by Fragile X) without having to take samples of living tissue from people who suffer from the disease.

These ESCs are also far superior to the cells of animals with diseases similar to those of humans (11). Verlinsky's method would have huge advantages if it really did work. Obtaining the large number of fresh human eggs needed for therapeutic cloning is not possible for legal reasons in many countries and is very expensive in countries where it is legal, such as the USA. By contrast, there is a virtually limitless supply of existing ESCs for fusion with adult cells. With Verlinsky's method, federally funded researchers could try to derive stembrids using already approved ESC lines. However, Verlinsky died recently, and the question is how quickly the research can go without him.

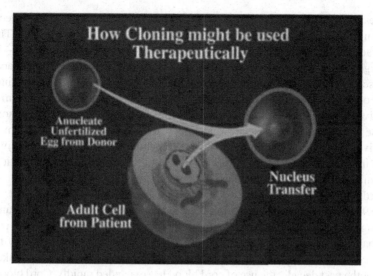

Fig. 16.1 Therapeutical cloning: Principles

The wind came from the other side when *Dr Stojkovic* (13, 14) from Serbia (after Dr. Hwang has withdrawn his discovery as not having meritorious values), working at New Castle, UK, then in Spain, and now again, in Serbia, Leskovac, has shown that analysis of arrested embryos demonstrated that these embryos express pluripotency marker genes such as OCT4, NANOG, and REX1. Derived hESC lines also expressed specific pluripotency markers (TRA-1-60, TRA-1-81, SSEA4, alkaline phosphatase, OCT4, NANOG, TERT, and REX1) and differentiated under in vitro and in vivo conditions into derivates of all three germ layers (14). All of the new lines, including lines derived from late arrested embryos, have had normal karyotypes. The line of work of this researcher and his group is actually the first one which has clearly and with no doubts demonstrated the possibility of establishing embryonic human cell line from arrested embryos. These results demonstrated that arrested embryos are additional valuable resources to surplus and donated developing embryos and should be used to study early human development or derive pluripotent hESC (14). The simplified scheme of therapeutic cloning is given in Fig. 16.1.

2. *Fetal stem cells*

As the embryo grows it accumulates additional embryonic stem cells in yolk sack. From weeks eight to 12 "fetal stem cells" are accumulated in the liver. Both embryonic and fetal stem cells generate the developing tissues and organs. At this stage such stem cells are designed as "multipotent" and they are more tissue specific rather than generating all of the body's 200 different cell types. Such stem cells are generally designated as "multipotent (15). However, some research suggests that at least some multipotent stem cells may be more plastic than first thought and may, under the right circumstances, become pluripotent. Up until week 12, fetal stem cells (as well as the embryonic stem cells which preceded them) can be transplanted into an individual, without being rejected. This is because they have little to none of the immune-triggering proteins—HLA antigens—on their surface (16). After the 12th week, fetal stem cells acquire these proteins, and they remain present on stem cells from this point on, including on adult stem cells. Thus, while some advocate therapeutic use of stem cells derived from cord blood, adult bone marrow, or the bloodstream, these sources pose the problem of possible rejection reactions (17). Therefore, stem cells derived from these sources may have therapeutic potential only when given to the individual from whom they were derived ("autologous" transplantation) or from an immunolgically matched donor ("allogenic" transplantation).

3. Cord blood stem cells

   In 1988, successful transplantation occurred in a young boy with Fanconi anemia using umbilical cord blood collected at the birth of his sibling (18). The patient remains alive and well to this date. In 1992, a patient was successfully transplanted with cord blood instead of bone marrow for the treatment of leukemia. Over the past decade, the use of cord blood has expanded rapidly. Cord blood has been used to transplant in any disease state for which bone marrow can be used in spite of some disadvantage—rather small number of cells collected in each unit; delayed engraftment of neutrophils is common with cord blood.

4. *Adult stem cells*

   Self-regeneration is the ability of stem cells to divide and produce more stem cells. During early development, the cell division is symmetrical, i.e., each cell divides to give rise to daughter cells, each with the same potential. Later in development, the cell divides asymmetrically with one of the daughter cells produced also a stem cell and the other a more differentiated cell. Here we meet the problem with cultivation of adult stem cells since most of the media applied induce early differentiation. Therefore, they were expanded in culture media enriched with cytokine cocktail, and assayed before and after the expansion either by cloning in vitro using limited dilution method in culture (19) or in vivo by transferring them through NOD/SCID mice (20) or adoptive humanized mouse model (21). All of these approaches have their advantages and disadvantages and still do not satisfy the optimal needs for expansion, HLA (MSH) compatibility, and consecutive simple therapeutic use of adult stem cells. For that reason, the mobilization of stem cells from intact bone marrow, for the purposes other than bone marrow transplantation (for example in the case of myocardial infarction or stroke as it was done by Orlic and Ratajczak in the first line), emerged as an idea to replace the in vitro manipulation, overcome in vitro expansion problems, and do the job in vivo. Unfortunately, as already mentioned, some individuals are poor mobilizers and some of them are irresponsive, for yet unknown reasons. And this is the limitation of otherwise splendidly designed clinical approach. Therefore, a lot of basic research is necessary to expand this field of stem cell application and with the better knowledge of the self-renewal control mechanism it will largely lead to efficient improvement, e.g., control and use of the fundamental principles for optimal expansion.

Focus of Stem Cell Research and Its Implication upon Neurological Stem Cell Therapy: Mapping the Molecular Scenario in Order to Engineer Repaired Brain Tissue

In order to be able to understand how to optimize the use of stem cells in clinical arena and brain tissue engineering scientists focused their research to the fundamentals—e.g., the chain of molecular events that regulate the decision of adult stem cells to perpetuate self-renewal or differentiate into more defined and/or mature cells.

Signaling Pathways that Maintain Stem Cells in an Undifferentiated State

The ability to maintain mammalian tissues throughout adult life depends on the persistence of stem cells. They are maintained in numerous adult tissues by self-renewal (dividing to make more stem cells), raising the question of whether this process is regulated by mechanisms that are conserved between tissues. A growing evidence that stem cells gone awry in their efforts to repair tissue damage could help explain why long-term irritation, such as from alcohol or heartburn, can create a breeding ground for certain cancers. It is too early to jump into generalization, but accumulated data could be summarized as follows:

- *Hedgehog and Wnt stem-cell repair mechanism*
  Two key chemical signals, called Hedgehog and Wnt ("wint") that are active in the stem cells that repair damaged tissue, have also recently and unexpectedly been found in certain hard-to-treat cancers, supporting an old idea that some cancers may start from normal stem cells that have somehow gone bad. Normal stem-cell self-renewal is a tightly regulated process, so the question is how and whether such regulation is circumvented in cancer.
- *Wnt*
  The place to start looking is the activity and regulation of Hedgehog and Wnt, which are best known for their roles in embryonic development, because recent studies show that they are key regulators of self-renewal in at least some of the body's normal tissue stem cells and are active in numerous cancer types (22–24). If these stem cells are the starting point of some cancers, multiple genetic and other changes may be required to trap the stem cell during chronic irritation, and perhaps many more changes to get the rapid growth of cancer. We need to figure out what those changes might be. Hedgehog activity has been found in certain cancers of the lung, brain, stomach, esophagus, skin, pancreas, bladder, muscle, and prostate (24). Similarly, Wnt activity has been tied to certain cancers of the colon, liver, blood, bone, and lung. In experiments at Hopkins and elsewhere, blocking Hedgehog and Wnt in laboratory-grown cancer cells and in animals has

been shown to kill cancer cells; so the pathways are potential targets for anticancer drugs (22, 23). The researchers caution that Hedgehog and Wnt blockers could affect normal processes that use these signals, including normal tissue repair, although short-term studies in mice have not yet found toxic side effects (23–25). Activation of the Wnt pathway is also associated with brain tumors. In particular, a subset of medulloblastomas has been reported to harbor mutations in *β-catenin, Axin,* or *APC* (26). Since these tumors have a distinct phenotype and molecular profile from medulloblastomas associated with hedgehog pathway mutations, they may arise from a distinct set of progenitors. To date, there have been no reports of Wnt pathway mutations in brain tumors outside the cerebellum. However, given the importance of Wnt signaling in self-renewal of progenitors throughout the brain, it is worth considering whether dysregulation of this pathway occurs in other types of brain tumors as well.

- *Other signaling pathways*
  The importance of these signaling pathways (JAK-STAT in embryonic stem cells, Notch, Jagg 1) in adult stem cells versus NSCs is under investigation. Stimulation of human epidermal differentiation by delta-notch signaling at the boundaries of stem-cell clusters has been found by two investigators, independently (27, 28). While the canonical Notch signaling pathway is deceptively simple, the consequences of the Notch activation on cell fate are complex and context dependent. The manner in which other signaling pathways cross-talk with Notch signaling appears to be extraordinarily complex. For example, the soluble Notch ligand, Delta-Fc, causes glial lineage determination by neural crest stem cells within 1 day in culture (28). One can hypothesize that these pathways could be involved in glial tumor growth in some instances. So, each type of brain tumor probably has different mechanisms of extensive growth and dedifferentiation. Therefore, this part of neurological stem cell therapy might be the most tricky to adjust since tumors vary from glioblastoma, meduloblastoma, and ependymoma to astrocytoma. It is impossible to expect general therapeutic approach for each of them. Therefore, the basic research in this field at molecular level is essential and the therapeutical management quite distinctive. Obviously, in this case the classical stem cell transplantation is not the answer.

Transcription Factors and Their Repressors

- *Bmi-1 dependence distinguishes neural stem cell self-renewal from progenitor proliferation*
  Bmi-1 regulates self-renewal of progenitors in the central and peripheral nervous systems (29), in part by repressing genes (such as p16Ink4a and p19Arf) that promote cellular senescence and cell death. In addition, *bmi-1* has recently been shown to be a target of the Shh signaling pathway and to be required for self-renewal of neural progenitors in the cerebellum (30). However, the polycomb family transcriptional repressor Bmi-1 is required for the self-renewal but not for the differentiation of stem cells in the hematopoietic system and peripheral and

central nervous systems. In each case, stem cells are formed in normal numbers during fetal development but exhibit impaired self-renewal and become depleted postnatally. Bmi-1 promotes stem cell self-renewal partly by repressing p16Ink4a, a cyclin-dependent kinase inhibitor, and p19Arf, a p53 agonist (29, 31). Both of these checkpoint proteins negatively regulate cell proliferation, and their increased expression has been associated with cellular senescence. This demonstrates that stem cells require mechanisms to prevent premature senescence in order to self-renew throughout adult life. In contrast, restricted neural progenitors from the enteric nervous system and forebrain proliferate normally in the absence of Bmi-1. Thus, Bmi-1 dependence is conserved between stem cells and distinguishes the cell cycle regulation of stem cells from the cell cycle regulation of at least some types of restricted progenitors. Using similar approaches, the additional pathways were studied that hypothetically will also regulate stem cell self-renewal, and that will contribute to understanding the molecular basis for self-renewal (29). Since stem cells persist throughout life by self-renewing in numerous tissues including the central and peripheral nervous system, this raises the issue of whether there is a conserved mechanism to effect self-renewing divisions. Deficiency in the polycomb family transcriptional repressor *Bmi-1* leads to progressive postnatal growth retardation and neurological defects. Weimanmn et al. have shown that *Bmi-1* is required for the self-renewal of stem cells in the peripheral and central nervous systems but not for their survival or differentiation (31). The reduced self-renewal of *Bmi-1*-deficient neural stem cells leads to their postnatal depletion. In the absence of *Bmi-1*, the cyclin-dependent kinase inhibitor gene *p16Ink4a* is upregulated in neural stem cells, reducing the rate of proliferation. Deficiency of *p16Ink4a* partially reverses the self-renewal defect in *Bmi-1−/−* neural stem cells (29, 31). This conserved requirement for *Bmi-1* to promote self-renewal and to repress *p16Ink4a* expression suggests that a common mechanism regulates the self-renewal and postnatal persistence of diverse types of stem cell. Restricted neural progenitors from the gut and forebrain proliferate normally in the absence of *Bmi-1*. Thus, *Bmi-1* dependence distinguishes stem cell self-renewal from restricted progenitor proliferation in these tissues. BM-1-deficient NSC exhibit a reduced rate of proliferation (31). Cells within *Bmi1−/−* stem cell colonies divide at a reduced rate, leading to reduced colony size, a reduced rate of BrdU incorporation into cells, and reduced stem cell self-renewal. Since *Bmi-1* is necessary for adult stem cells from the hematopoietic and nervous systems to self-renew normally, the defects in its expression or function might cause neurological problems.

Neurotropic Factors and Receptors

- Endothelin receptor B (Ednrb)
  Wild-type neural crest stem cells (NCSCs)
  NCSCs are one variety of tissue or "adult" stem cells. NCSCs persist in peripheral nerves throughout late gestation but their function is unknown (32).

The multilineage differentiation of NCSCs into glial and non-glial derivatives in the developing nerve appears to be regulated by neuregulin, notch ligands, and bone morphogenic proteins, as these factors are expressed in the developing nerve, and cause nerve NCSCs to generate Schwann cells and fibroblasts, but not neurons, in culture. Nerve development is thus more complex than was previously thought, involving NCSC self-renewal, lineage commitment, and multilineage differentiation. Wild-type NCSCs engraft and form neurons as efficiently in the aganglionic region of *Ednrb*-deficient gut as in wild-type gut, demonstrating that the *Ednrb*-deficient hindgut is permissive for the survival and differentiation of neural progenitors (33). This suggests a possible new strategy for treating Hirschsprung disease, a birth defect that is sometimes caused by mutations in *Ednrb* and that is associated with a failure to form enteric ganglia in the hindgut. *Ednrb* is required to modulate the response of NCSCs to migratory cues; loss of *Ednrb* leads to a failure of these cells to migrate into the hindgut, despite the fact that normal numbers of NCSCs are maintained throughout development in the proximal gut (34). The migration defect can be bypassed by transplanting NCSCs into the aganglionic region of the *Ednrb*-deficient embryonic gut. Transplanted alkaline phosphatase-expressing neural crest cells (34) that have engrafted in the hindgut wall of an *Ednrb*-deficient rat have demonstrated the ability to gain new insights into the etiology of disease as well as the ability to identify new potential therapeutic approaches by studying the regulation of stem cell function. The identification of NCSCs in the adult gut (enteric nervous system), with p75 (neurotrophin receptor), has been done in a transverse section of the postnatal rat gut. NCSCs were isolated from the adult gut by flow-cytometric purification of the cells that expressed the highest levels of p75. In this section, the cells that stained brightly for p75 localized to the myenteric and submucosal plexi of the enteric nervous system (34). The profound impact of these data is in the fact that prior to this work it was thought that NCSCs terminally differentiate during fetal development and do not persist in the adult peripheral nervous system.

- Glial cell-Derived Neurotrophic Factor (GDNF)
  An understanding of the mechanisms that regulate organogenesis from stem cells will make it possible to identify molecular links between stem cell function and disease (32–34). Combined gene expression profiling with reverse genetics and analyses of stem cell function in the hematopoietic and nervous systems to identify mechanisms that regulate organogenesis from stem cells and that lead to congenital disease when defective gave interesting results. Hirschsprung disease is a relatively common birth defect characterized by a failure to form enteric ganglia in the hindgut. It was found that it is caused by mutations in two pathways (the GDNF and endothelin signaling pathways, Ednrb pathway) that interact to regulate the generation and migration of NCSCs in the gut (32–35). Mutations in these pathways lead to a failure to form the nervous system in the hindgut by preventing NCSCs from migrating into the hindgut. These insights raise the possibility of treating this disease with stem

cell therapies that would bypass these defects. The studies described above suggest that brain tumor and neural stem cells resemble one another at a cellular and molecular level. Whether brain tumors actually arise from NSCs in most cases remains unclear, and a definitive answer may have to await the development of better markers for neural stem cells and progenitors. Nevertheless the fact that these cells express similar genes and show activation of similar signaling pathways suggests that they use similar strategies to achieve self-renewal and to generate heterogeneous populations of cells within a tissue. Understanding stem cell signaling pathways and developing ways to manipulate them will be critical in our effort to develop new treatments for tumors of the nervous system.

Potential Use of Stem Cells in Neurological and Other Diseases: Tissue Engineering at the Bench Site

Repair of damaged organ/tissue (myocardial, neuronal, liver, cartilage and bone, etc.) is the ultimate goal of stem cell therapy. The experimental and clinical trials have shown both in animal models and humans the neovascularization and myocardial tissue repair through transdifferentiation into myocardiocytes, or some other mechanisms (15, 35 and 39–45).

- *Organogenesis from stem cells*
  How do a small number of stem cells give rise to a complex three-dimensional tissue with different types of mature cells in different locations? This is the most fundamental question in organogenesis. The hematopoietic and nervous systems employ very different strategies for generating diversity from stem cells. The hematopoietic system assiduously avoids regional specialization by stem cells. Hematopoietic stem cells are distributed in different hematopoietic compartments throughout the body during fetal and adult life, and yet these spatially distinct stem cells do not exhibit intrinsic differences in the types of cells they generate (15, 46, 47). This contrasts with the nervous system, where even small differences in position are associated with the acquisition of different fates by stem cells.
  While local environmental differences play an important role in this generation of "neural diversity," we have found that intrinsic differences between stem cells are also critical. Part of the reason why different types of cells are generated in different regions of the nervous system is that intrinsically different types of stem cells are present in different regions of the nervous system. To understand the molecular basis for the regional patterning of neural stem cell function, we are studying how these differences are encoded.
- *Therapeutic implications for tissue-committed stem cells (TCSCs)*

To prove that the stem cells derived from BM and peripheral blood, including hematopoietic stem cells, are indeed transformed into solid-organ-specific cells, several conditions must be met:

1. The origin of the exogenous cell integrated into solid-organ time must be documented by cell marking, preferably at the single-cell level.
2. Cell should be processed with a minimum of "ex vivo" manipulation (e.g., culturing) which may make them more susceptible to crossing lineages.
3. The exogenous cells must be shown to have become an integral morphologic part of the newly acquired tissue.
4. Transformed cells must be shown to have acquired the function of the particular organ into which they have been integrated both by expressing organ-specific proteins and by showing specific organ function.

Organ/Tissue-specific niche (like in BM, liver, etc.) contains within the adult stem cells, which are circulating in a very low number in the blood. Accumulating evidence suggests that stem cells may also actively migrate/circulate in the post-natal period of life. Stem cell trafficking/circulation may be one of the crucial mechanisms that maintains the pool of stem cells dispersed in stem cell niches of the same tissue that are spread throughout different anatomical areas of the body. This phenomenon is very well described for HSC, but other TCSCs, for example, endothelial, skeletal muscle, skeletal, or neural stem cells, are probably circulating as well (46).

BM is the home of migrating stem cells with not only hematopoietic stem cells within their niches but also a small number of TCSCs, which might be the reason why many authors think that the HSC may transdifferentiate, although we do not have a direct proof for that. They might have plasticity, but not necessarily the "transdifferentional potential." What is differentiated in the tissue of injection might be TCSC characteristic for that tissue. It has been shown that the number of these cells is decreased with ageing (long-living and short-living mice and humans). It would be interesting to identify genes that are responsible for tissue distribution/expansion of TCSC. These genes could be involved in controlling the life span of the mammals. Therefore, BM stem cells are a heterogeneous population of cells with HSC and TCSC, the morphological and functional characteristics of which are different from those of HSC. Their number among BMMNC is very low (1 cell per 1,000–10,000 BMMNC) within young mammals and might play a role in small injuries. In severe injuries like heart infarct or stroke they have no possibility to reveal their full therapeutic potential (46). The allocation of these cells to the damaged areas depends on homing signals that may be inefficient in the presence of some other cytokines or proteolytic enzymes that are released from damaged tissue-associated leukocytes and macrophages. We can envision, for example, that metalloproteinases released from inflammatory cells may degrade SDF-1 locally, and thus perturb homing of CXCR4 + TCSC. There is possibility that these cells while "trapped" in BM are still in "dormant" stage—not fully functional and need the appropriate activation signals by unknown factors. These cells also, at least in some cases, could be attracted to the inflammatory areas, and if not properly incorporated into the damaged tissue they may transform and initiate tumor growth. These cells are by a deeper analytical approach categorized as very small embryonic like cells (VSELs) which are probably the cells found by other researchers, earlier in the past who were not able to reproduce them. Their presence in mouse and humans is confirmed and their characteristics given in Figs. 15.2 and 15.3.

Summary and Conclusions

Stem cells maintain tissues in adult life, and are regulated so that they generate exactly the right number of differentiating cells to balance cell loss from the tissue. Different stem cells from different organ sources seem to have different homing ability and capacity, but there is always a small fraction of them circulating throughout the body and trying to repair damaged tissues. Although stem cell research has been hailed for the potential to revolutionize the future of medicine with the ability to regenerate damaged and diseased organs, it has been highly controversial due to the ethical issues concerned with the culture and use of stem cells derived from human embryos. So, after all, looking critically at what we have today, the use of ESCs in therapy of different diseases is more problematic than the story that precedes their establishment. They are actually allogeneic transplants which need immunosupression due to in vitro intervention called stem cell genetic cloning, where embryonal cell is fused into hybrid with the nucleus of chosen egg cell in order to continue self-renewal and repair of the desired tissue line or organ (nuclear gene transfer technique). This is a critical issue from both scientific and clinical point of view, having in mind that the concept of VSEL (very small embryonic like adult, non-hematopoietic stem cells) has been recently developed and proved by Dr. Ratjczak's group (15). The question is whether ESCs should be used at all, if VSELs can already replace them, successfully. On the other hand, the number of VSELs is very low and requires mobilization to reach efficient level in the bloodstream in order to make efficient homing in targeted organ or tissue. Furthermore, some individuals are poor mobilizers and some cannot be mobilized at all. In those cases ESC could be of valuable help. A lot of further work and methodological advances are required, making the field of stem cell therapy more intriguing and closer to clarification of ideas and problem solutions in clinical application.

The unique characteristics of stem cells make them very promising potential for supplying cells and tissues instead of organs in a spectrum of devastating diseases from diabetes type1 to stroke, spinal cord injuries, and myocardial infarction (15, 35–47). In the situation when the number of people needing organ and tissue transplants exceeds the number of donated organs and tissues, this is the promise and hope, which deserve a deep and serious consideration.

References

1. Sell S (2004) Stem cells. In: Sell S, Sell S (eds) Stem cell handbook., pp 1–18
2. Forbes SJ, Vig P, Poulsom R, Wright NA, Alison MR (2002) Adult Stem Cell Plasticity: New Pathways of Tissue Regeneration become Visible. Clin Sci 103:355–369
3. Rando TA (2006) Stem cells, ageing and the quest for immortality. Nature online. Nature, 441:1080–1086. Accessed Feb 16 2006
4. Weinberg RA (1991) Tumor suppressor genes. Science 254(5035):1138–1146
5. Weissman I (2005) Stem cell research: paths to cancer therapies and regenerative medicine. JAMA 294(11):1359–66
6. Lasky JL, Liau LM (2006) Targeting stem cells in brain tumors. Technol Canc Res Treat 5(3):251–260, ISSN 1533-0346
7. Voskoboynik A, Simon-Blecher N, Soen Y, Rinkevich B, De Tomaso AW, Ishizuka KJ, Weissman IL (2007) Striving for normality: whole body regeneration through a series of abnormal generations. FASEB J 21(7):1335–44
8. Woo Suk Hwang (2004) Evidence of a pluripotent embryonic stem cell line derived from a cloned blastocyst. Science 303(5664):1669–1674
9. Hwang WS, Roh SI, Lee BC, Kang SK, Kwon DK, Kim S, Kim SJ, Park SW, Kwon HS, Lee CK, Lee JB, Kim JM, Ahn C, Paek SH, Chang SS, Koo JJ, Yoon HS, Hwang JH, Hwang YY, Park YS, Oh SK, Kim HS, Park JH, Moon SY, Schatten G (2005) Patient-specific embryonic stem cells derived from human SCNT blastocysts. Science 308:1777–1783
10. Verlinsky Y, Tur-Kaspa I, Cieslak J, Bernal A, Morris R, Taranissi M, Kaplan B, Kuliev A (2005) Preimplantation testing for chromosomal disorders improves reproductive outcome. Reprod Biomed Online 2(2):219–225
11. Daley, JK, Trounson A et al (2007) Ethics. The ISSCR guidelines for human embryonic stem cell research. Science 315(5812):603–604
12. Trounson A et al (2007) Ethics. The ISSCR guidelines for human embryonic stem cell research. Science 315(5812):603–604
13. Zhang X, Stojkovic P, Przyborski S, Cooke M, Armstrong L, Lako M, Stojkovic M (2006) Derivation of human embryonic stem cells from developing and arrested embryos. Stem Cells 24(12):2669–76
14. Ahmad S, Stewart R, Yung S, Kolli S, Armstrong L, Stojkovic M, Figueiredo F, Lako M (2007) Differentiation of human embryonic stem cells into corneal epithelial like cells by in vitro replication of the corneal epithelial stem cell niche. Stem Cells 25(5):1145–55
15. Kucia M, Reca R, Campbell FR, Surma-Zuba E, Majka M, Ratajczak M, Ratajczak MZ (2006) A population of very small embryonic–like (VSEL) CXR4+SSEA-1+Oct4+ stem cells identified in adult bone marrow. Leukemia 20:857–69
16. Guillot PV, Gotherstrom C, Chan J, Kurata H, Fisk NM (2007) Human first-trimester fetal MSC express pluripotency markers and grow faster and have longer telomeres than adult MSC. Stem Cells 25(3):646–54, Epub 2006 Nov 22
17. Guillot PV, O'Donoghue K, Kurata H, Fisk NM (2006) Fetal stem cells: betwixt and between. Semin Reprod Med 24(5):340–7, Review
18. Weiss ML, Troyer DL (2006) Stem cells in the umbilical cord. Stem Cell Rev 2(2):155–62, Review
19. Young-sup Y, Wecker A, Heyd L, Park JS, Losordo DW et al (2005) Clonally expanded novel multipotent stem cells from human bone marrow regenerate myocardium after myocar dial infarction. J Clin Invest 115(2):326–38
20. Ivanovic Z, Hermitte F, Brunet de la Grange P, Dazey B, Belloc F, Lacombe F, Vezon G, Praloran V (2004) Simultaneous maintenance of human cord blood SCID-repopulating cells and expansion of committed progenitors at low O_2 concentration (3%). Stem Cells 22:716–724
21. Traggiai E, Chicha L, Mazzucchelli L, Bronz L, Piffaretti JC, Lanzavecchia A, Manz MG (2004) Development of a human adaptive immune system in Cord Blood Cell-Transplanted Mice. Science 304(5667):104–7

22. Calabrese C, Poppleton H, Kocak M, Hogg TL, Fuller C, Hamner B, Young Oh E, Gaber MW, Finklestein D, Allen M, Frank A, Bayazitov IA, Zakharenko SS, Gajjar A, Davidoff A, Gilbertson RJ (2007) A perivascular niche for brain tumor stem cells. Cancer Cell 11(1):69–82

23. Lam L (2004) Beachy PA The Hedgehog response network: sensors, switches, and routers. Science 304(5678):1755–9

24. Karhadkar SS, Bova GS, Beachy PA et al (2004) Hedgehog signalling in prostate regeneration, neoplasia and metastasis. Nature 431(7009):707–12, Epub 2004 Sep 12

25. Bergman K, Graff GD (2007) The global stem cell patent landscape: implications for efficient technology transfer and commercial development. Nat Biotechnol 25(4):419–24

26. Wechsler-Reya RJ (2003) Recent Prog. Horm Res 58:249–261

27. Henrique D, Hirsonger E, Adam J, Le Roux I, Horowitz D, Lewis J (1997) Maintenance of neuroepithelial progenitor cells by Delta-Notch signalling in the embryonic chick retina. Curr Biol 10:491–500

28. Rohn JL, Lauring AS, Linenberger ML, Overbaugh J (1998) Transduction of Notch2 in feline leukemia virus-induced thymic lymphoma. Blood 92:1505–1511

29. Molofsky AV, Pardal R, Iwashita T, Park IK, Clarke MF, Morrison SJ (2003) Bmi-1 dependence distinguishes neural stem cell self-renewal from progenitor proliferation. Nature 425(6961):962–7, Epub 2003 Oct 22

30. Leung C, Lingbeek M, Shakhova O, Liu J, Tanger E, Saremaslani P, Van Lohuizen M, Marino S (2004) *Nature* 428:337–341. Abstract + References in Scopus | Cited By in Scopus

31. Weimann JM, Charlton CA, Brazelton TR, Hackman RC, Blau HM (2003) Contribution of transplanted bone marrow cells to Purkinje neurons in human adult brains. Proc Nat Acad Sci USA 100(4):2088–2093

32. Joseph NM, Mukouyama Y, Mosher JT, Jaegle M, Steven A, Crone SA, Dormand E-L, Lee KF, Meijer D, Anderson DJ, Morrison SJ (2004) Neural crest stem cells undergo multilineage differentiation in developing peripheral nerves to generate endoneurial fibroblasts in addition to Schwann cell. Development 131:5599–5612, Published by The Company of Biologists 2004

33. Iwashita et al (2003) Hirschsprung disease is linked to defects in neural crest stem cell function. Science 15 Aug: 972

34. Angrist M, Bolk S, Halushka M, Lapchak PA, Chakravarti A (1996) Germline mutations in glial cell line-derived neurotrophic factor (GDNF) and RET in a Hirschsprung disease patient. Nat Genet 14:341–344

35. Mezey E, Chandross KJ, Harta G, Mak RA, McKercher SR (2000) Turning blood into brain: cells bearing neuronal antigens generated in vivo from bone marrow. Science 290(5497): 1779–1782. doi:10.1126/science.290.5497.177

36. Brazelton TR, Rossi FM, Keshet GI, Blau HM (2000) From marrow to brain:expression of neuronal phenotypes in adult mice. Science 290(5497):1775–9

37. Chen J, Zhang ZG, Li Y, Wang L et al (2003) Intravenous administration of human bone marrow stromal cells induces angiogenesis in the ischemic boundary zone after stroke in rats. Circ Res 92(6):629–9

38. Bang OY, Lee JS, Lee PH et al (2005) Autologous mesenchymal stem cell. transplantation in stroke patients. Ann Neurol 57:874–882

39. Orlic D, Kajstura J, Chimenti S, Jakoniuk I, Anderson SM, Li B et al (2001) Bone marrow cells regenerate infarcted myocardium. Nature 410:701–5

40. Stamm C, Westphal B, Kleine HD, Petzsch M, Kittner C, Klinge H et al (2003) Autologous bone-marrow stem-cell transplantation for myocardial regeneration. Lancet 361:45–6

41. Strauer BE, Brehm M, Zeus T, Köstering M, Hernandez A, Sorg RV et al (2002) Repair of infarcted myocardium by autologous intracoronary mononuclear bone marrow cell transplantation in humans. Circulation 106:1913–8

42. Assmus B, Schachinger V, Teupe C, Britten M, Lehmann R, Döbert N et al (2002) Transplantation of progenitor cells and regeneration enhancement in acute myocardial infarction (TOPCARE-AMI). Circulation 106:3009–17

43. Tse HF, Kwong YM, Chan JKF, Lo G (2003) Ho C-L, Lau C-P. Angiogenesis in ischaemic myocardium by intramyocardial autologous bone marrow mononuclear cell implantation. Lancet 361:47–9
44. Perin EC, Dohmann HFR, Borojevic R, Silva SA, Sousa ALS, Mesquita CT et al (2003) Transendocardial, autologous bone marrow cell transplantation for severe, chronic ischemic heart failure. Circulation 107:2294–302
45. Obradović S, Rusović S, Dinčić D, Gligić B, Baškot B, Balint B, i sar (2003) Autologe pluripotentne progenitorne ćelije u lečenju ishemijske bolesti srca. Vojnosanit pregl 60(6):725–31
46. Kucia M, Ratajczak J, Ratajczak ZM (2005) Bone Marrow as a source of circulating CXR4+ tissue-commited Stem Cells. Biol Cell 97:133–146
47. Kucia M, Zhang YP, Reac R, Wysoczynski M, Machalinski B, Majka M, Ildstad ST, Ratajczak JU, Chields CB, Ratajczak MZ (2006) Cells enriched in markers of neural tissue-commited stem cells reside in the bone marrow and are mobilized into the peripheral blood following stroke. Leukemia 20:18–28

Chapter 17
Neurological Diseases and Stem Cell Therapy

In questions of science, the authority of a thousand is not worth the humble reasoning of a single individual.

Galileo Galilei

Can Stem Cell Therapy Work in Stroke? Progress from Animal Toward Human Studies

In stroke, occlusion of a cerebral artery leads to focal ischemia in a restricted central nervous system (CNS) region. Many different types of neurons and glial cells degenerate in stroke. It has not yet been convincingly demonstrated that neuronal replacement induces symptomatic relief in individuals who have suffered strokes. The approaches to regenerative stroke therapy vary at both experimental and clinical levels (Table 17.1).

These approaches vary from just a mobilization or transfusion of adult stem cells, across the use of embryonic stem (ES) cells, surgical implantation to the application of neurons derived from different stem cell sources [1–7]. It is convenient to think of the brain in evolutionary terms:

- The higher brain is concerned with intellect and advanced skills, such as movement and manual dexterity, as well as the reception of the senses.
- The function of the cerebellum is more primitive; it partially encircles the brain stem, symbolically suggesting its role as a stabilizer or moderator, smoothing movements and actions, and regulating the irregular impulses of the higher cortex into controlled activity, while maintaining balance and posture.

Between the cerebellum and the proximal extent of the spinal cord are the basal ganglia. These are amongst the most primitive areas of the brain and contain some of the basic regulatory centers for autonomic function as well as co-ordination.

M. Pavlovic and B. Balint, *Stem Cells and Tissue Engineering*,
SpringerBriefs in Electrical and Computer Engineering,
DOI 10.1007/978-1-4614-5505-9_17, © The Author(s) 2013

Table 17.1 Experimental designs and clinical trials that have inspired neurological stem-cell research in humans

Disease category	Model	Stem-cell source	Route of cell application	Outcome	Study
Neurodegenerative diseases	Experimental BMT	BM cells	Intraperitoneal	Generation of cells expressing neuronal markers	
	Experimental BMT	BM cells	Intravenous	Generation of cells expressing neuronal markers	
	Clinical BMT (allogeneric)	BM cells	Intravenous	Generation of cells expressing neuronal markers	(Y-chromosome determined by FISH on female samples)
	Clinical BMT (allogeneric)	BM cells	Intravenous	Possible formation of Purkinje neurons	
Middle cerebral artery occlusion (stroke)	Experimental cord blood trans-plantation	Stem-cells derived from umbilical cord blood	Intravenous	Improved functional recovery from neurological deficit	
	Clinical mesenchymal BMSC autologous transplant	MSCs transplantation	Intravenous	Improved functional recovery	
	Experimental mice B16, 129	Neural TCSCs	In vitro mobilization chemoattraction experiment	Chemoattraction of neural TCSCs to supernatant from damaged brain tissue	

"Experimental" denotes data drawn from nonhuman settings

BMT bone marrow transplantation

Thus, in the only reported clinical trial in persons with stroke affected basal ganglia, received implants of neurons generated from the human NT-2 teratocarcinoma cell line into the infracted area. The introduction of a proliferation step as free-floating cell spheres cuts the total time needed to obtain high yields of purified NT-2 neurons to about 24–28 days [8]. The cells obtained show neuronal morphology and migrate to form ganglion-like cell conglomerates. Differentiated cells express neuronal polarity markers such as the cytoskeleton-associated proteins MAP2 and Tau. Moreover, the generation of neurons in sphere cultures induced immunoreactivity to the ELAV-like neuronal RNA-binding proteins HuC/D, which have been implicated in mechanisms of nerve cell differentiation [8]. However, there is no substantial new neuron formation in the cerebral cortex after stroke. Whether the new neurons formed after stroke are functional is unknown.

Recent findings in rodents suggest an alternative approach to cell therapy in stroke based upon self-repair [9]. Stem cells taken from adult human bone marrow have been manipulated by scientists at the Maxine Dunitz Neurosurgical Institute at Cedars-Sinai Medical Center to generate aggregates of cells called spheres that are similar to those derived from neural stem cells of the brain [10]. In addition, the bone marrow-derived adult stem cells, which could be differentiated into neurons and other cells making up the CNS, spread far and wide and behaved like neural stem cells when transplanted into the brain tissue of chicken embryos [10]. Results of the experiments, described in the February 2007 of the Journal of Neuroscience Research, support the concept of using adult bone marrow-derived stem cells to create therapies to treat brain tumors, strokes, and neurodegenerative diseases [10]. Similar study using bone marrow-derived stem cells of rats appeared as the cover article of the December 2002 issue of *Experimental Neurology* [11–14]. These findings reinforce the data that came from the study of rat (adult) bone marrow-derived stem cells. Using two methods, scientists showed evidence for the bone marrow derived stem cells being neural cells, and demonstrated that it is feasible to grow the cells in large numbers. They also documented that these cells function electrophysiologically as neurons, using similar voltage-regulating mechanisms [11].

Progressing from the rat study to experiments with human cells and transplantation into mammal brain tissue, the international research team continues to build a foundation for translating laboratory research into human clinical trials. Based on some studies to date, a patient's own bone marrow appears to offer a viable and renewable source of neural stem cells, allowing us to avoid many of the issues related to other types of stem cells [15–17]. The replacement of damaged brain cells with healthy cells cultured from stem cells is considered to potentially be a promising therapy for the treatment of stroke, neurodegenerative disorders, and even brain tumors, but finding a reliable source for generating neural cells for transplantation has been a challenge. The use of embryonic and fetal tissue has raised ethical questions among some and brings with it the possibility of immune rejection. And while neural stem cells can be taken from brain tissue, the removal of healthy tissue from a patient's brain introduces a new set of safety, practicality, and ethical issues.

Mobilization as a Scientifically Based Novel Approach to Stroke Therapy

In the studies performed by Shyu and Lee with collaborators (2005 and 2006) on rats, the mobilization step was used to approach stem cell therapy and functional recovery of stroke [18, 19]. Subcutaneous administration of G-CSF one day after right middle cerebral artery ligation and consecutive ischemia in rats, in the dosage of 50 µg/kg/day during 5 days, has improved body asymmetry and locomotor activity in experimental group of rats compared to control which did not receive G-CSF [18]. The elevated body swing test was used to estimate body asymmetry. Locomotor activity was assessed in activity chamber. Beside these results in the same study, the authors have found a decreased infarction volume, determined by MRI from an average of 176 mm^3 in saline-treated controls (ligation only) to 61 mm^3 in G-CSF-treated animals. BrdU labeling of injected cells used to follow the engraftment of G-CSF-mobilized hematopoietic stem cells (HSC) has shown that BrdU immunoreactive cells have been found in the ipsilateral cortex near the infracted boundary and subventricular region. It is noteworthy to mention that BrdU immunoreactive cells were also found around the lumen of varying calibers of blood vessels in the perivascular portion (also on the endothelial cell lining of the vessel wall). The results of double-staining immunohistochemistry under laser scanning confocal microscopy have shown some BrdU cells co-localized for antibodies for Neu-N, MAP-2, GFAP, and VWF in the brain of G-CSF-treated rats, indicating differentiation of dividing stem cells into their progenitors from HSCs. The increased expression of CXCR4 was also found in the G-CSF-treated rats in comparison with contralateral side and control rats, as well as their presence in cortical and vascular endothelial cells indicating that cerebroendothelial SDF-1 can be a chemoattractant for peripheral blood stem cells. The data were encouraging in terms of further consideration; that disruption of the blood–brain barrier may facilitate selective entry of HSCs into the ischemic rather than the non-ischemic contralateral hemisphere. It is suggested by others that in ischemic rat brain a number of neurotrophic factors are released, which have been shown to result in human bone marrow stromal cell factor production [19]. Therefore, the authors speculated that ischemic damage to brain tissue may result in the release of trophic factors, which in turn may target HSCs to damaged tissue. It would be of great importance to determine which signaling molecules attract HSCs and direct their migration to damaged areas.

The other study performed by the same authors, quite recently, not using mobilization but transplantation of PBMCs [18] has shown by approach with laser scanning confocal microscopy that rats receiving intracerebral PBMCs transplantation were seen to differentiate into glial cells GFAP+, Neu-N+cells, and vWF+positive cells, thereby enhancing neuroplastic effects in the ischemic brain. In addition, PBSC implantation promoted the formation of new vessels, thereby increasing the local cortical blood flow in the alchemic hemispheres. They suggested that involvement of stem cell delivered macrophage-microglial cells, and beta-1 integrin expression might be enhancing factors of this angiogenic architecture over the ischemic brain. Further studies are needed to prove this and other hypotheses.

In the recent work performed by Ratajczak and his group [20–26], with labeled TCSCs and their mobilization into peripheral blood following stroke, the authors hypothesized that the postnatal BM harbors a non-hematopoietic population of cells that express markers on neural TCSCs that may account for the beneficial effects of BM-derived cells in neural regeneration. These, neural TCSCs were chemoattracted to the damaged neural tissue in an:

- SDF-1–
- CXCR4–
- HGF-c-Met–
- LIF-LIF-R-dependent manner

These cells not only express neural lineage markers such as:

- Beta3-tubulin
- Nestin
- NeuN
- GFAPbut more importantly, form neurospheres in vitro. So, the BM has been shown to contain a mobile pool of endothelial TCSCs that may play an important role in organ regeneration.

Thus, the study was performed on B16, 129 wild-type mice and in vitro, for neurosphere formation. The same group has also detected embryonic stem-cell derived microvesicles that are selectively highly enriched in mRNA for several pluripotent transcription factors as compared to parental ES [25]. These vesicles suggest an evidence for horizontal transfer of mRNA and protein delivery to target cells where they may be translated into the corresponding proteins [25]. The concept that BM-derived cells participate in neural regeneration remains highly controversial and the identity of the specific cell types involved remains unknown. The results of Ratajczak's group (by using FACS analysis combined with analysis of neural markers at the mRNA and protein levels) although still at the experimental level revealed at least in mouse, the presence of this mobile cellular pool residing in non-hematopoietic CXCR 4+/Sca–/lin–/CD45– BM mononuclear cell fraction [26]. Neural TCSC are mobilized into the peripheral blood during the stroke and chemoattracted to the damaged neural tissue in a manner described above. Very recently, the pool of VSELs and tissue committed progenitors was detected also in humans [26]. Therefore, it would be worthwhile trying this product in clinical arena. In summary, there is a potential for adult stem cells to regenerate damaged area in human brain after the stroke, but the entire treatment has to be finely tuned with respect to type and location of damage, and type of the cells that would be used. This apparently involves optimal time and route of administration and optimizations of all conditions in the patient that would allow for such approach.

Can Stem Cell Therapy Work in Neurological Tumors?

In their recent work, the Cedars-Sinai researchers documented that several genes that speed up and control the proliferation process could be used to rapidly expand the supply of marrow-derived neural stem cells, writing in the article that "this novel method of expansion may prove to be useful in the design of novel therapeutics for the treatment of brain disorders, including tumors" [27].

What Regulates Cancer Stem Cells?

Based on our previous large-scale gene expression study by cDNA microarray, upregulation of one gene, called Maternal Embryonic Leucine-zipper Kinase (MELK), was found both in normal neural stem cell (NSC) and in malignant BT [28, 29]. Therefore, it was likely that MELK plays a role in the similar subset of cells existing in both NSC and BT, which is the "stem cell" in normal brains and in malignant tumors. So far, by using gene knockdown technique, called RNA interference, scientists identified that MELK is a critical regulator of normal NSC proliferation. They have initiated treatment of MELK RNA interference of BT cells [29].

Role of Stem Cells in Brain Cancer

Ontogeny (development of an organism) and oncology (cancer development) share many common features. From the 1870s the connection between development and cancer has been reported for various types of cancers. Existence of "cancer stem cells" with aberrant cell division has also been reported more recently. The connection between cancer and development is clearly evident in teratocarcinomas. The teratocarcinomas are able to differentiate into normal mature cells when transplanted into another animal [30, 31]. This alternation between developmental and tumor cells status demonstrates how closely development and cancer are related. McCulloch explored the connection between normal development of blood cells and leukemia [32]. According to him, normal hematopoietic development requires the interaction of stem cell factor with its receptor, c-kit [32]. A hierarchy of stem and progenitor cells differentiates and produces different sublineages of cells resulting from response to varied growth factors. Malignancies of the hematopoietic system originate from two sources: those with an increased growth in an early stem cell produce acute leukemia, while those that arise from a decreased response to death or differentiation in a stem cell produce chronic leukemia.

The present-day challenge is to decode the common molecular mechanism and genes involved in self-renewal for cancer cells and stem cells.

The notion that antiangiogenic therapy targets cancer stem cells has important implications for evaluating and optimizing the use of antiangiogenic drugs in cancer.

Scientists are currently dissecting the various cellular and protein components of the niche to identify the critical parts that maintain cancer stem cells [27]. This may help identify new therapeutic targets. Yang and Wechsler-Reya announced that this work "highlights the importance of the vascular microenvironment in brain tumor growth" [33]. Brain cancer stem cells are maintained within vascular niches. Although these cells may be resistant to conventional treatment, preliminary studies suggest that antiangiogenic drugs can block tumor growth. Cancers share more properties of normal developing tissues than we may have appreciated. This work opens up new avenues for treatments, but suggests also that we need to work hard to define truly how cancers and normal tissues differ [27].

Gilbertson and associates investigated whether brain tumor stem cells develop within niche microenvironments in the brain vasculature. The majority of cancer stem cells in brain tumors were closely associated with tumor capillaries [27]. In vitro, brain tumor stem cells associated rapidly with endothelial vascular tubes, forming close contacts along their lengths, the results indicate [32]. Further experiments showed that endothelial cells maintained self-renewing and undifferentiated brain tumor cells and promoted the propagation of brain tumors in vivo. Depletion of brain tumor blood vessels effectively eradicated the population of self-renewing tumor cells, the report indicates. This led investigators to propose "that antiangiogenic drugs arrest brain tumor growth, at least in part, by disrupting a vascular niche microenvironment that is critical for the maintenance of cancer stem cells" [33].

By recent advance in fundamental knowledge about NSCs and brain tumors (BT), a small fraction of tumor cells within malignant BT has been found to show strikingly similar characteristics with normal NSC. Interestingly, only this subset of tumor cells, called "cancer stem cells," can initiate the entire malignant tumor and further cause metastasis to other organs, while the rest of a large heterogeneous population in BT does not. This "stem cell" in tumor was first identified in acute myeloblastic leukemia with significant similarity to normal HSC, and the breast cancer was the first solid cancer, which was found to contain stem cell fraction. With regard to BT, two independent studies including group led by Prof. Harley I. Kornblum, Pharmacology, UCLA have shed light on the stem cell population in some of pediatric malignant tumors in 2003 [34]. Several studies from other groups have succeeded in reproduction of the existence of the "brain tumor stem cells" since then [35].

Clinical Significance of Identification of Cancer Stem Cells

After the isolation of stem cells in BT, scientists have started to identify the critical regulator genes of the stem cells in BT (http://www.who.int/cardiovascular_diseases/resources/atlas/en/) [36–42]. This project was based on one hypothesis; by blocking the critical regulator of brain tumor stem cells, they should be able to stop the malignant tumors to proliferate, keep growing, and invading into the adjacent normal brain structures, and eventually, kill the patients with malignant tumors.

Alzheimer's Disease and Stem Cell Therapy

Alzheimer's disease (AD) is a type of dementia due to formation of amyloid plaques in the brain tissue. More than 500,000 people in Britain for example, suffer from Alzheimer's disease, for which there is currently no cure. This figure is expected to increase dramatically in the coming years as the population ages. Therefore, the advance in adult stem cell regenerative therapy of this type of dementia would open the perspective to a broad range and high number of patients.

Researchers at the University of California in Irvine successfully used injections of neural stem cells to repair damaged brain cells [43]. Although the experiment was only done on mice, the researchers are confident that the technique may one day be used on humans to restore memory lost during the late stages of Alzheimer's. Doctor Frank LaFerla, Director of the University's Institute for Memory Impairments and Neurological Disorders, said there is "a lot of hope" that the findings could lead to "a useful treatment for Alzheimer's" [43]. A research team led by University of Central Florida professor Kiminobu Sugaya has discovered a compound related to DNA (staurosporine) that could improve the results of stem cell treatments for Alzheimer's patients [44–51]. The research team found that treating bone marrow cells with the compound made adult stem cells more likely to turn into brain cells in experiments with rats.

The granulocyte-colony stimulating factor (G-CSF) is a blood stem cell growth factor or hormone routinely administered to cancer patients whose blood stem cells and white blood cells have been depleted following chemotherapy or radiation. G-CSF stimulates the bone marrow to produce more white blood cells needed to fight infection. It is also used to boost the numbers of stem cells circulating in the blood of donors before the cells are harvested for bone marrow transplants. Advanced clinical trials are now investigating the effectiveness of G-CSF to treat stroke, and the compound was safe and well tolerated in early clinical studies of ischemic stroke patients. G-CSF has been used and studied clinically for a long time, but this is the first group to apply it to Alzheimer's disease [52]. This growth factor could potentially provide a powerful new therapy for Alzheimer's disease—one that may actually reverse disease, not just alleviate symptoms like currently available drugs. The researchers showed that injections under the skin of filgrastim (Neupogen®)

- One of the three commercially available G-CSF compounds.
- Mobilized blood stem cells in the bone marrow and neural stem cells within the brain and both of these actions led to improved memory and learning behavior in the Alzheimer's mice on the basis of reactive microglia derived from stem cells that are destroying deposits of amyloid plaques in brain tissue.

So far, a human growth factor that stimulates blood stem cells to proliferate in the bone marrow reverses memory impairment in mice genetically altered to develop Alzheimer's disease [52]. The G-CSF significantly reduced levels of the brain-clogging protein beta-amyloid deposited in excess in the brains of the Alzheimer's mice increased the production of new neurons and promoted nerve cell connections. The beauty in this less invasive approach is that it obviates the need for neurosurgery to transplant stem cells into the brain. Based on the promising findings in mice, the Alzheimer's Drug Discovery Foundation is funding a pilot clinical trial at USF's Byrd Alzheimer's Center. The randomized, controlled trial, led by Dr. Sanchez-Ramos and Dr. Ashok Raj, will test the safety and effectiveness of filgrastim in 12 patients with mild-to-moderate Alzheimer's disease.

Parkinson's Disease and Stem Cell Therapy

Parkinson's disease is the second most common neurodegenerative disease following Alzheimer's. Approximately 1.5 million people in the USA suffer from Parkinson's disease, which is caused when 80% or more of dopamine-producing neurons in the substantia nigra of the brain die. Normally, dopamine is secreted from the substantia nigra and transmitted to another part of the midbrain. This allows body movements to be smooth and coordinated.

The main pathology in PD is degeneration of nigrostriatal dopaminergic neurons. The predominant neuropathological feature of Parkinson's disease is characterized by a loss of the neuromelanin-containing dopaminergic cells of the substantia nigra. It seems, however, not the primary loss of these cells themselves that is the fundamental pathology, but the consequent loss of their neurotransmitter, dopamine. The eventual result is a degeneration of the complete dopaminergic nigrostriatal pathway, although it is estimated that 80% of dopamine in the striatum is lost before symptoms become clinically apparent (http://www.parkinson.org/site/pp. asp?c=9dJFJLPwB&b=71125). At post-mortem, physiological deterioration is also observed in areas of the brainstem such as the nuclei of the locus coeruleus and the dorsal motor nucleus of the vagus. Lewy bodies (eosinophilic intracytoplasmic inclusions) are found in the brainstem and other parts of the brain, and are diagnostic of the condition, at least at post-mortem.

Patients with Parkinson's disease are treated with the drug levodopa (or L-dopa), which is converted to dopamine in the body. Initially effective, the treatment's success is reduced over time and side effects increase, leaving the patient helpless (http://stemcells.nih.gov/info/scireport/chapter8.asp). It has been recognized that dopamine-producing cells are required to reverse Parkinson's disease. Since the 1970s, many types of dopamine-producing cells have been used for transplantation. These include adrenal glands from the patient, human fetal tissue, and fetal tissue from pigs [53]. Limited success has been achieved with these cells. Rat and monkey models of Parkinson's were used to test fetal mesencephalic cells [54, 55]. Success with animal models led to clinical trials.

Trials with human fetal mesencephalic tissue reach in postmitotic dopaminergic neurons have provided a proof of principle that neuronal replacement can work in human brain. They reversed the impairment of cortical activation underlying akinesia. Several open-label trials have reported clinical benefits [54, 55]. Unfortunately, dyskinesias can develop after transplantation and become troublesome in 7–15% patients [56, 57]. These adverse effects are not due to dopaminergic overgrowth, but rather by uneven and patchy reinnervation or by chronic inflammatory and immune responses around the graft. Only 5–10% of fetal mesencephalic grafts are dopaminergic neurons. It is not yet known whether it is favorable to implant a pure population of dopaminergic neurons or whether the graft should also contain a specific composition of other neuron types and glial cells to induce maximum symptomatic relief.

Fetal tissue transplantation has been performed in 350 patients, including trials using pig fetal tissue. So far, the success of reversing Parkinson's disease using fetal tissue has been limited at best. However, in the most successful cases, patients have been able to lead an independent life without L-dopa treatment (http://www.parkinson. org/site/pp.asp?c=9dJFJLPwB&b=71125). The limitations include (a) lack of sufficient tissue for the number of patients in need, (b) variation in results between patients ranging from no benefit to reversal of symptoms, and (c) Occurrence of uncontrolled flailing movements (called dyskinesias).

The many criteria for the cells used in therapy include the ability to produce dopamine, to divide and survive in the brain, and to integrate into the host brain. For these reasons, differentiated embryonic stem cells offer more promise. Mouse ES cells have been used in rat models of Parkinson's disease and recently human ES cells have been reported to differentiate into dopamine-producing neurons in culture [56, 57].

Another consideration is the immune problem. It was believed that the brain is an immunologically privileged site tolerating transplanted cells from a different individual (meaning that the immune system will not attack tissue transplanted into this location). However, a recent report challenges this view (http://stemcells.nih. gov/info/scireport/chapter8.asp) [54]. For this reason autologous cells may offer a safer alternative. Neural stem cells and HSC are both likely candidates [54]. Also, dental pulp cells in both rats and humans produce neurotrophic factors and are a candidate for autologous transplantation in Parkinson's [58].

If the adult stem cell treatment cannot help as the regenerative approach to PD, maybe we shall have to use embryonic stem cell (ESC). Somatic cell nuclear transfer was used to clone Dolly, the sheep. Since then, seven animal species have been cloned using this technique. A modified version for use in human's ES is as follows: The patient's DNA is injected into an enucleated unfertilized egg and used to generate ES cells which are then cultured and allowed to differentiate, followed by transplantation into the patient. This technique is called therapeutic cloning. The use of such cells may bypass the ethical objections and immunological issues of using ES cells and is the future of stem-cell clinical application.

Nevertheless, very recently the encouraging results have been published on the use of adult stem cells in one patient suffering PD, which caused improvement in entire clinical picture and put the perspective of this kind of treatment into the scope of regenerative stem cell medicine. UCLA researchers published their results in February issue of the *Bentham Open Stem Cell Journal* which outlines the long-term results of the trial (Dr. Michel Levesque). They have documented the first successful adult neural stem cell transplantation to reverse the effects of

Parkinson's disease and demonstrated the long-term safety and therapeutic effects of this approach.

Dr. Levesque's team was able to isolate patient-derived neural stem cells, multiply them in vitro, and ultimately differentiate them to produce mature neurons before they are reintroduced into the brain. The team was able to inject the adult stem cells without the need for immunosuppressants.

Unlike embryonic stem cells, adult stem cell injections do not cause a patient's immune system to reject the cells. These adult stem cells were highly beneficial for the patient involved in the study.

Of particular note are the striking results this study yielded—for the 5 years following the procedure the patient's motor scales improved by over 80% for at least 36 months. A larger clinical trial would replicate the findings. While the data show that the technique needs refinement, this patient went for several years with little to no symptoms of this disease, even with only half of the brain treated with his own adult stem cells [57].

Stem Cell Transplant Research, Parkinson's Disease

Future will show whether the results will continue to prove that adult stem cells outpace their embryonic counterparts. Does it mean that we need to take notice that it is not embryonic stem cells that provide promise of treatments in the future, but rather it is adult stem cells that are already providing safe and effective therapies for patients now, without the problems of rejection or tumors? The future work will show, since we can learn just from our work (Fig. 17.1).

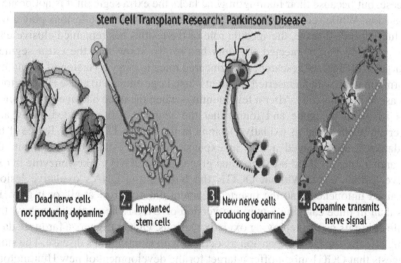

Fig. 17.1 Stem cell transplant research, Parkinson's disease

Huntington's Disease and Stem Cell Therapy

It is a progressive, fatal autosomal dominant neurodegenerative disorder caused by increased CAG repeats in the Huntington gene. Clinically it is characterized by chorea and progressive dementia-deterioration in cognitive and neuropsychiatric functions [59–76]. The main pathology is the loss of medium spiny projection neurons in the striatum, due to mutation in the huntingtin gene. Cell therapy in Huntington's disease aims at restoring brain function by replacing these neurons [64, 75, 76]. There is another aspect to this disease which is an inherited neurodegenerative disorder that affects roughly 30,000 Americans, incurable and fatal. But a new discovery about how cells repair their DNA points to a possible way to stop or slow the onset of the disease gives the hope that the cure might be found. Unlike most inherited diseases, Huntington's disease symptoms usually do not appear until middle age, leading scientists to wonder what triggers the disease onset, and whether it can be halted—or at least slowed down.

People with the disease have a version of a gene called huntingtin that carries an extra segment with a particular sequence of repeated subunits. If the segment is too large, the gene produces a faulty protein that has a destructive effect in the brain. Huntington's disease is a progressive disease, but nobody knows exactly why. Research work supports the idea that the disease progresses when the extra segment expands over time in non-dividing cells such as nerve cells. McMurray's study shows that the inserted segment grows when cells try to remove oxidative lesions, which are caused by by-products of the oxygen we breathe [62]. DNA repair enzymes initially keep oxidative lesions in check, but over time, increasing numbers of lesions overwhelm the repair systems.

Oxidative lesions also accumulate in people who do not have Huntington's disease, but because their huntingtin gene lacks the extra segment it is not prone to expansion. While scientists have long suspected that oxidative lesions play a role in Huntington's disease, the specific role of the lesions has remained elusive until now. Nobody has connected the dots before. To show that the extra segment enlarges with age, the researchers engineered mice to carry a version of the human huntingtin gene with an inserted segment—one large enough to cause Huntington's disease in humans [62]. After a few months—when the mice had aged—the scientists analyzed the gene and found that the segment had expanded. They also observed an increase in oxidative lesions in the mouse DNA [62]. To see if the oxidative lesions played a role in expansion of the extra DNA segment, the researchers next deleted 8-oxoguanine glycosylase (OGG1), a key enzyme in oxidative lesion repair. Without OGG1, the bulk of the DNA's oxidative lesions remained untouched, and the inserted segment did not grow at all, or it grew far less than in mice carrying a working version of OGG1. These findings show that while doing its part in removing oxidative lesions, OGG1 triggers a far more damaging effect—the DNA expansion associated with Huntington's disease. The study suggests that OGG1 might offer a target for the development of new Huntington's disease treatments. McMurray's team is already pursuing this path and is screening for small molecules that block OGG1 function. This work may also be relevant to research on other diseases, such as Alzheimer's and Parkinson's, in which oxidative lesions are believed to play a role.

In animal models, intrastriatal grafts of fetal striatal tissue containing projection neurons reestablish connections with the globus palidus and receive inputs from host cerebral cortex. This level of reconstruction of corticostriatopallidal circuitry is sufficient to reverse motor and cognitive deficits in rats and monkeys [62].

Clinical trials with intrastriatal transplantation of human fetal striatal tissue support the cell replacement strategy in Huntington's disease (HD). The grafts survived without typical pathology, contained striate projection neurons, and received afferents from the patient's brain. However, the extent of clinical benefits was unclear [54, 55]. One open-label trial indicated cognitive and motor improvements [76] whereas outcome was unchanged in other. Clinical improvement was associated with reduction of striatal and cortical metabolism, suggesting that the grafts had restored function in striato-cortical neural loops. Substantial benefit following cell therapy will require that many more grafted striatal neurons survive than the low numbers achieved in the trials with the fetal tissue. The stem cell technology could markedly increase the availability of such cells.

Basic research should now explore how to generate and select striatal projection neurons from stem cells, and probably combine stem cell therapy with accommodation of oxidative damages in DNA involved in this disease.

Researchers are using the grid-computation method, sharing tasks over multiple computers, in order to learn about the molecular nature of diseases that are caused from errors in protein folding. The aim of this project is to better understand protein folding, what happens when proteins do not fold correctly, and the diseases that result from this protein misfolding (http://www.aboutdementia.com/articles/about-huntingtons/huntingtons-causes.php). Once the causes of protein misfolding are discovered, cures for diseases such as Alzheimer's disease, Huntington's disease, cystic fibrosis, BSE, and cancers could find a cure. Misfolding of proteins causes clumps of proteins to gather in the brain, causing diseases. Proteins fold very quickly, some as fast as a millionth of a second. One of the diseases, currently focusing on, is Huntington's disease. This disease results from the aggregation of various proteins. If proteins contain long enough strands that contain numerous repeats of the amino acid, glutamine aggregates begin to form, causing the disease. This abnormally long of a repeated section of the amino acid glutamine in the DNA sequence has been thought to stem from an inherited gene and has been given the name, huntingtin gene. A healthy person has a string of 9–39 glutamines, whereas, Huntington's patients have 36–121 glutamines (http://www.aboutdementia.com/articles/about-huntingtons/huntingtons-causes. php). This protein folding aspect is examining the structure of poly-glutamine aggregates and is attempting to predict the pathway that forms these aggregates. The goal of these Huntington's studies is to find drug design approaches so Huntington's can be find a method to stop the disease from forming the aggregate pathway. Thus, it can be used in future as a preventive, while stem cell replacement therapy, as a curative approach to disease.

Results of a new study published online on August 10, 2009 in the Proceedings of the National Academy of Sciences question the long-term effects of transplanted cells in the brains of patients suffering from Huntington's disease. The study, conducted by Dr. Francesca Cicchetti of Laval University in Québec, Canada, Dr. Thomas B. Freeman of the University of South Florida (USF) Department of Neurosurgery and Brain Repair, Tampa, FL, and colleagues, provides the first demonstration that transplanted cells fail to offer a long-term replacement for degenerating neurons in patients with Huntington's disease.

Amyotrophic Lateral Sclerosis and Stem-Cell Therapy

Amyotrophic lateral sclerosis (ALS), also known as Lou Gehrig's disease, is a progressive disease that manifests as a gradual evolution and spread of weakness and wasting of the affected patient's muscles. It leads to dysfunction, disability, and ultimately death or chronic ventilator dependency, which occur an average of 3 years after weakness is first detected [77]. The pathophysiology of ALS is complex and poorly understood despite many years of study, and because of this, no treatments have proved effective in slowing disease progression to any significant degree. Some scientists have suggested that stem cells, with their ability to differentiate into a variety of cell lines, could play an important role in restoring damaged motor neurons and even generating new ones [77]. Are stem cells a dream come true for those with ALS? Or are they something less? The answer to both questions may be "yes." With the lack of effective drug treatments for ALS, and compelling preclinical data, stem-cell research has highlighted this disease as a candidate for stem-cell treatment [78, 79]. Stem-cell transplantation is an attractive strategy for neurological diseases and early successes in animal models of neurodegenerative disease generated optimism about restoring function or delaying degeneration in human beings.

The restricted potential of adult stem cells has been challenged over the past several years by reports on their ability to acquire new unexpected fates beyond their embryonic lineage (transdifferentiation). Therefore, autologous or allogeneic stem cells, undifferentiated or transdifferentiated and manipulated epigenetically or genetically, could be a candidate source for local or systemic cell-therapies in ALS. Albert Clement and colleagues showed that in SOD1G93A chimeric mice, motorneuron degeneration requires damage from mutant SOD1 acting in non-neuronal cells. Wild-type non-neuronal (glial) cells could delay degeneration and extend survival of mutant-expressing motorneurons [70]. Letizia Mazzini and colleagues [77] injected autologous bone marrow-derived stem cells into the spinal cord of seven ALS patients. These investigators reported that the procedure had a reasonable margin of clinical safety. The question was: where next? In 2004, the same group of authors published the new data [79]. After the results they got with animal models of ALS that stem cells significantly slow the progression of the disease and prolong survival, they have evaluated the feasibility and safety of a method of intraspinal cord implantation of autologous mesenchymal stem cells (MSCs) in a few well-monitored patients with ALS. For the treatment, bone marrow collection was performed according to the standard procedure by aspiration from the posterior iliac crest. Ex vivo expansion of MSC was induced according to Pittenger's protocol. The cells were suspended in 2 ml of autologous cerebrospinal fluid and transplanted into the spinal cord by a micrometric pump injector. No patient manifested major adverse events such as respiratory failure or death [79]. Minor adverse events were intercostal pain irradiation (four patients) which was reversible after a mean period of 3 days after surgery, and leg sensory dysesthesia (five patients) which was reversible after a mean period of 6 weeks after surgery. No modification of the spinal cord volume or other signs of abnormal cell proliferation were observed. They concluded that results appear to demonstrate that the procedures of ex vivo expansion of autologous MSCs and of transplantation into the spinal cord of humans are safe and well tolerated by ALS patients.

The success of cell-replacement therapy in ALS will depend a lot on preclinical evidence, because of the complexity and precision of the pattern of connectivity that needs to be restored in degenerating motorneurons. It is probably that adult stem-cell therapy will need to be used with other drugs or treatments, such as antioxidants and/or infusion of trophic molecules.

Cambria Biosciences, LLC is the recipient of a grant from the ALS Association supporting research to identify new drug candidates for the treatment of ALS. The research initiative, Translational Research Advancing Therapy for ALS (TREAT ALS), aims to accelerate the process of moving good ideas from the research arena into clinical trials and then patients.

Cellular Transplantation Strategies for Spinal Cord Injury

The spinal cord connects the brain with the rest of the body by sending out millions of electrical signals. When there is injury to the spinal cord and this connection breaks paralysis occurs. As of now, spinal cord damage is irreversible, leaving approximately 250,000 Americans in a devastating position. However, in the past 10 years, as stem cell research has developed, new exciting possibilities have arisen for people suffering from spinal cord injury. Crush injuries to the spinal cord are common, as opposed to clean cuts that sever the cord without any damage to overlying tissue and bone. Crush injuries usually result in low blood flow to the affected area, producing an ischemic condition. Fluid buildup often results, leading to compression from swelling and secondary ischemia. This damage can exceed the amount produced from the primary injury. A whole host of toxic conditions lead to the production of molecules that recruit glial cells to infiltrate the site in an effort to affect repair; unfortunately however, poor design of the system leads to reactive gliosis, which basically means the glial cells are looking to "plug the hole" instead of forming a nice, stable tube to guide the damaged axon back to its target. This glial scarring is an enormous barrier to regeneration of the spinal cord and recovery of function.

Stem cells represent a viable treatment option following spinal cord injury. Undifferentiated stem cells excrete a variety of neurotrophic factors that encourage axon growth, promote the replacement of damaged non-neural structures such as blood vessels, promote the breakdown of the glial scar, and temper inflammatory responses. Embryonic stem cells in particular have a penchant for adopting the glial phenotype, that is they will readily transform into the support cells required by neurons (e.g., astrocytes, oligodendrocytes) once they are transfused into the site of injury. They may also be used to overcome glial repulsion of axons; myelinating cells produce inhibitory factors that can prevent an axon from regenerating. With that in mind, there are a couple exciting papers coming out. The first [80] points to uses embryonic stem cells in the rat. These cells, when added to the site of damage along with a PDE-4 inhibitor to block the axon-repulsive effects of glia, were experimentally differentiated into a *neural* phenotype to form bridge connections between the degenerating axons and the muscle. The interesting manipulation in this paper was the infusion of cells that produce the trophic factor GDNF into the target muscle; GDNF provides a signal that attracts growth of axons from the embryonic stem cells.

A second paper [81] expands upon the promise of animal stem cell models of spinal cord injury by using human neural stem cells in a rodent model. They demonstrate differentiation of the human cells into neuronal and glial tissue, axon remyelination, synapse formation, and locomotor recovery. It seems, then, that stem cell therapies hold promise for treatment of traumatic spinal cord injury. While much work remains to develop a stable, consistent model in animals we are definitely making progress and a variety of very creative approaches are being used. Some of these approaches point directly to potential of human adult and/or embryonic stem cells [81–83]. A truly pro-life culture would embrace the exploration and use of these technologies for the benefit of all its citizens. One strategy to repair a damaged spinal cord involves stimulation of axon regrowth in order to reestablish the broken connection. When the dendrites receive information the cell body generates a nerve impulse, which travels along the axon to another "target" cell. At the target cell—a muscle cell, another nerve cell, or gland cell—the axon divides into a multitude of nerve endings. The tip of each of these endings is called the axon terminal and located very close to the target cell [80, 81]. Here the axon forms a synapse allowing neurotransmitters to travel across a small gap (25 nm wide) and fuse with the receptors of the next cell—this is how electric signals are sent from the brain to other parts of the body. When neurons die, connection between axon terminals and receptors is broken and the CNS can no longer function.

There are over 100 billion neurons in the CNS and as many as 10,000 different subtypes of these neurons [82]. The incredible power of the brain to process information exists in the massive amount of neurons and synapses. Neurons are not the only cell in the CNS—glial cells also exist in even greater numbers than neurons. These cells come in different forms with a variety of different functions, all helping the CNS to operate. Two glial cells related to the spinal cord are oligodendrocytes and astrocytes. Oligodendrocytes are responsible for producing myelin, a fatty substance that provides electrical insulation on the axons. Myelin allows electric signals to be sent at a rate of 100 m/s as opposed to 1 m/s without myelin [83]. Death of oligodendrocytes results in demyelination, halting communication between the brain and the rest of the body. Astrocytes break down and remove harmful proteins, as well as secrete proteins called neurotrophic factors, which help neurons survive and grow. Astrocytes also respond to injury: they clear away debris, an action resulting in formation of glial scarring.

The spinal cord needs more protection than any other organ or system because unlike other organs, the spinal cord cannot repair itself [80]. The complex interactions between the brain and neurons, in combination with the enormous number of individual neurons and synapses, make reconnection of the nerve cells extremely difficult. The spinal column supplies the main defense of the spinal cord, providing a protective barrier against injury. The syrinx, a fluid filled area, offers additional protection by absorbing shock. Unfortunately, both of these defenses cause complications upon injury. Swelling causes additional damage to the spinal cord as pressure builds in the confined space between the cord and vertebrae. The syrinx contributes to scar tissue that builds up around the area of injury. Scar tissue blocks the neurons from reconnecting once the cord has been severed.

Often the cord is not completely severed during injury; even so, swelling cuts off the blood supply to the neurons and glial cells. Without a blood supply these cells die. Additional cell death occurs as cells from the immune system migrate to the injury site. In order for a connection to be reestablished new neurons and glial cells must regenerate to replace the injured ones. Up until about 10 years ago people believed that there was no possibility for neurogenesis of adult nerve cells. Once nerve cells were damaged they were gone, eliminating hope for complete recovery from paralysis. As a result, treatments for spinal cord injury focused on prevention of further damage (secondary damage) and rehabilitation. While the majority of cells found in the CNS are born during the embryonic and early postnatal period, scientists like Raynolds and Weiss, even not so recently discovered that new neurons are continuously added to two specific regions of the adult mammalian brain [84]. Neural stem cells were isolated from the dentate gyrus of the hippocampus and the walls of the ventricular system called the ependymal layer. The progeny of these stem cells differentiate in the granule cell layer, meaning neurogenesis continues late into adult rodent life. These stem cells also migrate along the rostral migratory stream to the olfactory bulb, where they differentiate into neurons and glial cells [85]. Nerve cell differentiation has been witnessed in vivo, as well as in vitro when stimulated with an epidermal growth factor [86]. The discovery of differentiating stem cells in the brain revolutionized the way scientists think about treating spinal cord injury. Suddenly the chance for partial or possibly full recovery from paralysis seemed like a plausible option, based upon this "broken dogma." Attention shifted to regenerating the neurons and glial cells as a solution to spinal cord injury.

Along with pluripotent stem cells progenitor cells, a more restricted type of stem cells are found in the hippocampus and ependymal layer. These cells are immature cells that are predetermined to differentiate into neurons, oligodendrocytes, and astrocytes. In 1995 Frissen observed that the presence of nestin increases in response to spinal cord injury and it has been repeated [87]. Nestin is a protein expressed by stem cells: presence of it indicates neural stem cells are much more active then previously believed. Our brain naturally increases the production of stem cells to aid an injured CNS. If the brain responds in this way, why does not the spinal cord repair itself? In 1999, Johansson and Momma observed that the only active progenitor cells were differentiating into astrocytes [88]. They labeled ependymal cells with a Dil injection so migration could be followed. After making lesions in the spinal cord they waited 4 weeks and then observed the progress of the ependymal cells. They tested the cells found in the scar tissue around the site of injury and found that all DIL marked cells were astrocytes. This indicates that the progeny from ependymal cells had only differentiated to astrocytes. Stem cells do respond to spinal cord injury, just not for the purpose of reestablishing connection between neurons.

This realization sparked scientist's interest in understanding what triggers these progenitor cells to proliferate. The active progenitor cells may be ineffective in maintaining a functional CNS after injury, but if scientists could learn what signals triggered differentiation, perhaps they could induce differentiation of neurons and oligodendrocytes. Scientists began to focus on neurotrophic factors that triggered this differentiation, specifically the presence of brain-derived neurotrophic factors (BDNF) and neurotrophin 3 and 4 (NT-3 and NT-4). In the early 1990s these trophic factors were targeted as what triggered axon growth during early development. NT-3 also is expressed in greater amounts in response to spinal cord injury. In 1994 Schwab reported dramatic increase in function, and regrowth of a partially severed cord of rats after treatment with NT-3 [89]. In 1997 Grill, Gage, and colleagues published a paper examining the effects of transplanted NT-3 on motor skills and morphology after induced spinal injury in mice [90]. They focused on the corticospinal tract, the pathway in charge of making voluntary movements. NT-3 has been previously observed to promote regrowth of corticospinal axons and preserves degenerating motor neurons.

Grill and colleagues induced lesions in the dorsal hemisection of adult rat's spinal cord, resulting in severely limited motor ability. Next grafts of syngenic fibroblasts, genetically altered to produce NT-3, were transplanted into the lesion cavity of the experimental group. These rats were kept alive for 3 months and put though a series of tests to monitor motor improvement. These tests examined coordination, ability to walk on inclined surfaces and precision of foot placement. After 3 months these rats were killed for the purpose of a quantitative cell count.

Recipients of the NT-3 secreting grafts showed significant improvement in motor skills over the control group, although they did not recover to the full ability they had before injury. After 3 months recipients of the NT-3 grafts demonstrated growth of corticospinal axons up to 8 mm from where the transplant had taken place. Only the injured axons at the lesion site showed any sign of regrowth [90]. Uninjured axons showed no effort to reestablish connections across the site of injury [90]. This suggests that NT-3 only responds when corticospinal axons are injured. If scientists could pinpoint signals triggering this response, there is potential to manipulate the process in a manner causing neural cells to differentiate.

Triggering neurotrophic factors in hopes of inducing progenitors to proliferate is one of the two major areas of study in spinal cord regeneration. Scientists also can derive undifferentiated embryonic stem cells (ES cells) from fetal spinal cord tissue and then mature them into cells that are suitable to implant into the damaged spinal cord [16, 17]. When using ES cells, researchers have two options: they can treat ES cells, allowing them to mature into CNS cells in vitro before transplantation, or they can directly implant differentiated cells and depend on signals from the brain mature the cells. This technique became possible when Reynolds and Weiss found that stem cells taken from the brain could be propagated in vitro. This allowed labs to duplicate what occurs naturally in the brain, and attempt to use the product to regrow the damaged cells.

In December of 1999 McDonald and colleagues from Washington University School of medicine successfully implanted ES cells in laboratory rats. McDonald induced thoracic spinal cord injury in rats using a metal rod 2.5 mm in diameter resulting in paralysis [80]. Nine days after the injury McDonald and colleagues transplanted roughly 1 million ES embryoid bodies pretreated with retinoic acid into the syrinx that had formed around the contusion. During the 9 days that passed between injury and transplantation, all the standard events following a spinal cord injury occurred. At the time of injury some cells died immediately, followed by a second wave of apoptosis within the first 24 h [80]. The center of the bruised spine filled with fluid becoming a cyst referred to as syrinx. McDonald injected the ES cells into this cavity. Two weeks after the transplantation ES stem cells filled the area normally occupied by glial scarring. After 5 weeks the stem cells had migrated further away from the implantation site. Although a number of them had died, there was still enough for the rats to have a growing supply of neurons and glial cells. Most of the surviving cells were oligodendrocytes and astrocytes, but some neurons were found in the middle of the cord. The rats regained limited use of their legs. Paralysis had been cured.

McDonalds work in 1999 represented new successes in stem cell technology, but there are still many years of work ahead of us before any of this technology can be tested in humans. A major obstacle remains: although scientists are achieving results, they do not understand the factors responsible for what occurs. In McDonalds study, the regaining of functions could result from the few differentiated neurons. Another possibility could be that the high differentiation of oligodendrocytes remyelinated enough axons to reestablish communication. Or perhaps functions regained due to ES cells producing growth factors—more research will have to be done before these options are narrowed down. Additional to unclear understanding of the process, other complications exist. Any introduction of foreign cells into the body triggers the immune system. ES cells would not simply be accepted into the host CNS. McDonald used cyclosporine to prevent rejection in the rats, but things get more complicated when testing begins on humans. The brain and spinal cord are complex, mysterious realms of the body—until science can predict the exact affect of evolving technologies, no testing on humans can occur.

A major motivation behind spinal cord research has been Christopher Reeve [91]. Injured in a horseback riding incident, Christopher Reeve suffered a cervical spinal cord injury that left him quadriplegic. Thus, he began the Christopher Reeve Paralysis Foundation (CRPF). CPRF funds research to treat or cure paralysis resulting from spinal cord injury or other CNS disorders. CPRF supports a Research

Consortium, which collaborates the work of nine laboratories, as well as funds an international individual grants program. Several of the labs involved in the Research consortium focus on stem cells, making a lot of progress. The Salk Institute, run by Dr. Fred Gage, examines the progenitor cells differentiating into glial cells. Someday they hope to manipulate these progenitor cells, inducing differentiation into neural cells.

There are a lot of people who find stem cell research extremely unethical. Scientists have found the most success with ES cells taken from embryoid spinal cords: although the ES cells are taken from embryos consisting at most of 64 cells, they still have the potential to develop into a human being. People who believe life begins at conception remain morally against stem cell research. Justification is that the stem cells are derived from embryos discarded from fertility clinics. These embryos would be wasted if not used for stem cell research. Spinal cord injury research represents a new and rising field—more progress has been made in the last 5 years than in the previous 50. This sudden success resulted from the new understanding of stem cell technology. The concept of stem cells has gotten a lot of press lately—from Time Magazine to the television show South Park. The realization that stem cells have the potential to differentiate into neural cells opens new doors, destroying the accepted idea that adult neurogenesis is not an option. With these new possibilities, stem cell research has evolved into an exciting new field. There is a lot of room to grow—who knows what new discoveries the future will bring. The difficulty with treating spinal cord injuries arises from a number of factors. Firstly there is the primary damage to the axons of the spinal cord itself, resulting in mechanical damage that can inhibit neurotransmission and transport of cellular material to and from the distal cord. The damaged cord must also compensate for secondary damage such as the generation of free radicals, a lack of oxygen to the affected area (anoxia), glial scarring, and a host of other issues. The future work is necessary to overcome these particular limitations.

Summary, Conclusions, and Perspectives

Remarkable progress has been achieved in studying stem cells. Adult stem cells, taken from sources such as bone marrow and cord blood, have now been successfully used to treat well over 70 medical conditions. Last year a woman was given a new windpipe grown from her own stem cells in an advance which scientists said heralded "a new age in surgical care." Despite advances using adult stem cells, the Human Fertilization and Embryology Act was passed last year liberalizing the law governing scientists use of the controversial embryonic stem cells.

However, many scientists consider this branch of research, in which human embryos are destroyed by the process of harvesting stem cells, to be far less promising. In 2008 scientist Colin McGuckin left his position as professor of regenerative medicine at Newcastle University because the Government was failing to fund adult stem cell research.

The most exciting use of cultured stem cells is the promise for curing many devastating diseases like Parkinson's and other neurological illnesses. However, more basic research remains before stem-cell-based therapy is widely used. Development of stem cell-based therapies for neurodegenerative disorders is still at an early stage. Many basic issues remain to be resolved, and we need to move forward with caution and avoid scientifically ill-founded trials in affected individuals. One challenge now is to identify molecular determinants of stem cell proliferation so as to control undesired growth and genetic alterations of ESCs. We also need to know how to pattern adult stem cells to obtain a more complete repertoire of various types of cells for replacement, and how to induce effective functional integration of the stem cell-derived neurons into existing neural and synaptic network. Technological advances will be needed to make precise genetic modifications of stem cells or their progeny that will enhance their capacity for migration, integration, and pathway reconstruction.

The potential of the brain self-repair mechanisms is virtually unexplored. We need to develop technologies for genetic labeling of stem cell progeny so that we can firmly establish where neurogenesis occurs and which cell types are generated following damage. The functional properties of the new neurons and their ability to form appropriate afferent and efferent connections should be determined. We also need to identify, with the aid of genomic and proteomic approaches, the cellular and molecular players that, in a concerted action, regulate different steps of neurogenesis. On the basis of this knowledge, we should design strategies to deliver molecules that improve the yield of new functional neurons and other cells in the damaged area.

The use of embryonic stem cells (which can evolve into any type of cell in the body) has been surrounded by controversy. The essential importance of these findings are in that if other scientists can duplicate the process on a larger scale, it could reduce the need for embryonic stem cells in research and eliminate rejection problems associated with using stem cells from an outside donor. Researchers worldwide share opinion that various types of stem cells hold great promise for understanding and treating a wide variety of diseases. In a discovery that has the potential to change the face of stem cell research, a University of Louisville scientist has identified VSELs in the adult body that seem to behave like embryonic stem cells. Yet, of the stem cells discussed in this review ES cells are still considered to have the most capacity to differentiate into a variety of cells and their proliferation capacity is also unsurpassed by any other cell type. There are three major problems with ES cells; ethical issues, immunological rejection problems (that potentially could be solved by therapeutical cloning), and the potential of developing teratomas.

In the future, ideally, adult (somatic) stem cells from the patient will be extracted and manipulated, and then reintroduced into the same patient to cure debilitating diseases, including neurological. This would preclude the use of embryonic stem cells for cell therapy, eliminate the ethical objections against stem cell research, and also resolve immunological rejection problems. However, at present, the cell proliferation and differentiation potential of embryonic stem cells due to all obstacles mentioned above remains still more likely to produce a cure than do the adult (somatic) cells.

References

1. Pluchino S, Zanotti L, Deleldi M, Martino G (2005) Neural stem cells and their use as therapeutic tool in neurological disorders. Curr Drug Target 6(1):3–19
2. Ben-Hur T, Einstein O, Bulte JWM (2005) Stem cell therapy for myelin diseases. Curr Pharm Des 11(10):1255–1265
3. Barker RA, Jain M, Armstrong RJE, Caldwell MA (2003) Stem cells and neurological disease. J Neurol Neurosurg Psychiatr 74:553–557
4. Herzog EL, Chai L, Krause DS (2003) Plasticity of marrow derived stem cells. Blood 102(10):3483–3493
5. Long Y, Yang KY (2003) Bone marrow derived cells for brain repair: recent findings and current controversies. Curr Mol Med 3(8):719–725
6. Song SJ, Sanzhez-Ramos J (2003) Brain as the sea of marrow. Exp Neurol 184(1):54–60
7. Priller J (2003) Adult bone marrow cells populate the brain. Histochem Cell Biol 120(2):85–89
8. Hara K, Yasuhara T, Maki M, Matsukawa N, Masuda Seong T, Yu J, Ali M, Yu G, Xu Seung L, DavidU K, Hess and Cesar C, Borlongan V (2008) Neural progenitor NT2N cell lines from teratocarcinoma for transplantation therapy in stroke. Prog Neurobiol 85(3):318–334
9. Darsalia V, Kallur T, Kokaia Z (2007) Survival, migration and neuronal differentiation of human fetal striatal and cortical neural stem cells grafted in stroke-damaged rat striatum. Europ J Neurosci 26(3):605–614
10. Molina-Holgado F, Rubio-Araiz A, Garcia-Ovejero D, Moore WRJ, Arevalo-Martin A, Gomez-Torres O, Molina-Holgado E (2007) CB2 cannabinoid receptors promot mouse neural stem cell proliferation. Europ J Neurosci 25:3
11. Ma V, Fitzgerald W, Liu Q-Y, Shaugnessy TJ, Maric D, Lin HJ, Alkon DL, Barker JL (2004) CNS stem and progenitor cell differentiation into functional neuronal circuits in three dimensional collagen gels. Exp Neur 190(2):276–288
12. Hess DC, Hill WD, Carroll JE, Borlongan CV (2004) Do bone marrow cells generate neurons? Arch Neurol 61(4):483–485
13. Vitry S, Bertrand JY, Cumano A, Dubois-Dalcq M (2003) Primordial hematopoietic stem cells generate microglia but not myelin-forming cells in a neural environment. J Neurosci 23(33):10724–10731
14. Seaberg RM, Smukler SR, Kieeffer TJ, Enikolopov G, Asghar Z, Wheeler MB, Korbutt G, van der Kooy D (2004) Clonal identification of multipotent precursors from adult mouse pancreas that generate neural and pancreatic lineages. Nat Biotechnol 22:1115–1124
15. Johansson CB, Momma S, Clarke DL, Risling M, Lendahl U, Frisen J (1999) Identification of a neural stem cell in the adult mammalian central nervous system. Cell 96:25–34
16. Gage FH, Kempermann G, Palmer TD, Peterson DA, Jasodhara R (1998) Multipotent progenitor cells in the adult dentate gyrus. J Neurobiol 36(2):249–266
17. Gage FH, Coates PW, Palmer TD, Kuhn HG, Fisher LJ, Suhonen JO, Peterson DA, Suhr ST, Jasodhara R (1995) Survival and differentiation of adult neuronal progenitor cells transplanted to the adult brain. PNAS 92(25):11879–11883
18. Shyu W-C, Lin S-Z, Lee C-C, Liu DD, Li H (2006) Granulocyte colony stimulating factor for acute ischemic stroke: a randomized controlled trial. CMAJ 174:927–933
19. Lee ST, Chu K, Jung KH, Ko SY, Kim EH, Sinn DI, Lee YS, Lo EH, Kim M, Roh JK (2005) Granulocyte colony stimulating factor enhances angiogenesis after focal cerebral ischemia. Brain Res 1058:120–128
20. Kucia M, Reca R, Jala VR, Dawn B, Ratajczak J, Ratajczak MZ (2005) Bone marrow as home of heterogeneous populations of nonhematopoietic stem cells. Leukemia 19:1118–1127
21. Kucia M, Ratajczak J, Ratajczak ZM (2005) Bone marrow as a source of circulating CXR4+ tissue-committed stem cells. Biol Cell 97:133–146
22. Kucia M, Ratajczak J, Ratjczak MX (2005) Are bone marrow cells plastic or heterogeneous—that is the question. Exp Hematol 33(6):613–623

23. Ratajczak J, Miekus K, Kucia M, Zhang J, Reca R, Dvorak P, Ratajczak MZ (2006) Embryonic stem cell-derived microvesicles reprogram hematopoietic progenitors: evidence for horizontal transfer of mRNA and protein delivery. Leukemia 20:847–856

24. Kucia M, Wojakowski W, Ryan R, Machalinski B, Gozdzik J, Majka M, Baran J, Ratajczak J, Ratjczak MZ (2006) The migration of bone marrow-derived non-hematopoietic tissue-committed stem cells is regulated in and SDF-1-, HGF-, and LIF dependent manner. Arch Immunol Ther Exp 54(2):121–135

25. Kucia M, Zhang YP, Reac R, Wysoczynski M, Machalinski B, Majka M, Ildstad ST, Ratajczak JU, Chields CB, Ratajczak MZ (2006) Cells enriched in markers of neural tissue-committed stem cells reside in the bone marrow and are mobilized into the peripheral blood following stroke. Leukemia 20:18–28

26. Kucia M, Reca R, Campbell FR, Surma-Zuba E, Majka M, Ratajczak M, Ratajczak MZ (2006) A population of very small embryonic-like (VSEL) CXR4+SSEA-1+Oct4+ stem cells identified in adult bone marrow. Leukemia 20:857–869

27. Gilbertson RJ (2007) Brain tumor stem cells lurk in perivascular niches. Cancer Cell 11: 3–5

28. Nakano I, Dougherty JD, Kim K, Geschwind DH, Kornblum HI (2007) Phosphoserine phosphatase is expressed in neural stem cell niche and regulates neural stem cell proliferation. Stem Cells 25(8):1975–1984

29. Nakano I, Masterman-Smith M, Horvath S, Paucar AA, Lilievre V, Waschek JA, Lazareff JA, Freije WA, Gilbertson RJ, Liau LM, Geschwind DH, Nelson S, Mischel PS, Kornblum HI (2007) Maternal embryonic leucine zipper kinase (MELK) is a key regulator of the proliferation of malignant brain tumors, including brain tumor stem cells. J Neurosci Res 86(1):48–60

30. Virchow R (1863) Cellular pathology as based upon physiological and pathological histology. Lippincott, Philadelphia

31. Stevens LC (1970) Experimental production of testicular teratomas in mice of strains 129, A/He, and their F1hybrids. J Natl Cancer Inst 44:923–929

32. McCulloch EA, Minden MD, Miyauchi J, Kelleher CA, Wang C (1988) Stem cell renewal and differentiation in acute myeloblastic leukaemia. J Cell Sci Suppl 10:267–281

33. Yang Z-J, Ellis T, Markant SL, Read T-A, Kessler JD, Bourboulas M, Schüller U, Machold R, Fishell G, Rowitch DH, Wainwright BH, Wechsler-Reya RJ (2008) Medulloblastoma can be initiated in lineage-restricted progenitors or stem cells. Cancer Cell 14:135–145

34. Hemmati HD, Nakano I, Lazareff JA, Masterman-Smith M, Geschwind DH, Bronner-Fraser M, Kornblum HI (2003) Cancerous stem cells can arise from pediatric brain tumors. PNAS 100(25):15178–15183

35. Nakano I, Kornblum HI (2009) Methods for analysis of brain tumor stem cell and neural stem cell self-renewal. Methods Mol Biol 568:37–56

36. Till JE, McCulloch EA (1961) A direct measurement of the radiation sensitivity of normal mouse bone marrow cells. Radiat Res 14:2213–2222

37. Thomas ED (1999) Bone marrow transplantation: a review. Semin Hematol 36:95–103

38. Jackson KA, Goodell MA (2004) Generation and stem cell repair of cardiac tissue. In: Sell S (ed) Stem cell handbook. Humana Press, Totowa, pp 259–266

39. Barker RA, Widner H (2004) Immune problems in the central nervous system cell therapy. NeuroRx 1:472–481

40. Espinosa-Heidmann DG, Caicedo A, Hernandez EP, Csaky KG, Cousins SW (2003) Bone marrow-derived progenitor cells contribute to experimental choroidal neovascularization. Invest Ophthalmol Vis Sci 44(11):4914–4919

41. Locatelli F, Corti S, Donadoni C, Guglieri M, Capra F et al (2003) Neuronal differentiation of murine bone marrow Thy-1- and Sca 1-positiev cells. J Hematother Stem Cell Res 12(6):727–734

42. Jin HK, Schuchman EH (2003) Ex vivo gene therapy using bone marrow-derived cells: combined effects of intracerebral and intravenous transplantation in a mouse model of Niemann-Pick disease. Mol Ther 8(6):876–885

43. Kitazawa M, Vasilevko V, Cribbs DH, LaFerla FM (2009) Immunization with amyloid-beta attebuates inclusion body myositis-like myopathy and motor impairment in a transgenic mouse. J Neurosc 29(19):6132–6141
44. Sugaya K (2003) Stem cell therapy—a new option for AD patients? In: Richter RW, Richter B (eds) Alzheimer's disease—the basics. A Physician's guide to the practical management. Humana Press Spring, Totowa
45. Sugaya K (2005) Stem cell strategies for Alzheimer's disease. In: Hanin I, Cacabelos R, Fisher A (eds) Recent progress in Alzheimer's and Parkinson's diseases. Taylor & Francis, London, pp 183–190
46. Pulido JS, Sugaya K (2005) Papel de las celulas madre en la degeneracion macular asociada a l'edad. In: Mones J, Gomez-Ula (eds) Degeneracion macular asocial a la edad. Prous Science, Barcelona, pp 343–350
47. Sugaya K (2005) Possible stem cell therapy for Alzheimer's disease and its future direction. In: Kanazawa I, Shibazaki H, Tougi H (eds) Modern medical treatment in neurology. Brains Network, Japan, pp 108–114
48. Kwak Y-D, Sugaya K (2006) RNA interference in human NTera-2/D1 cell lines using human U6 promoter-based siRNA PCR products. Biotechnology and Bioprocess Engineering 2006, 11(3):273–276
49. Kwak Y-D, Choumkina E, Sugaya K (2006) Amyloid precursor protein is involved in staurosporine induced glial differentiation of neural progenitor cells. Biochem Biophys Res Commun 344(1):431–437
50. Kwak Y-D, Kim HM, Qu T, Brannen CL, Soba P, Majumdar A, Kaplan A, Beyreuther K, Sugaya K (2006) Amyloid precursor protein cause glial differentiation of human neural stem cell. Stem Cell Dev 15:381–389
51. Sugaya K, Qu K, Sugaya T, Pappas GD (2006) Genetically engineered human mesenchymal stem cells produce Met-Enkephalin at augmented higher levels in vitro. Cell Transplant 15:225–230
52. Sanchez-Ramos, Raj A (2009) Blood stem cell growth factor reverses memory decline in mice. The randomized, controlled trial. At: www.physorg.com/newsn165684042.html. Accessed 6 Jun 2012
53. Baier PC, Schindehutte HJ, Thinane K, Flugge G, Fuchs E, Mansouri A, Paulus W, Gruss P, Trenwalder C (2004) Behavioral changes in unilaterally 6-hydroxy-dopamine lesioned rats after transplantation of differentiated mouse embryonic stem cells without morphological integration. Stem Cells 22:396–404
54. Lindvall O, Bjorklund A (2004) Cell therapy in Parkinson's disease. NeuroRx 1:382–393
55. Polgar S, Morris ME, Reilly S, Bilney B, Sanberg PR (2003) Reconstructive neurosurgery for Parkinson's disease: a systematic review and preliminary meta-analysis. Brain Res Bull 60: 1–24
56. Zheng X, Cai J, Chen J, Luo Y, Zhi-Bing Y, Fotter E, Wang Y, Harvey B, Miura T, Backman C, Chen G-J, Rao MS, Freed WJ (2004) Dopaminergic differentiation of human embryonic stem cells. Stem Cells 22:925–940
57. Peterson DA (2004) Stem cell therapy for neurological disease and injury. Panminerva Med 46(1):75–80
58. Todorovic V, Markovic D, Milosevic-Jovcic N, Petakov M, Balint B, Colic M, Milenkovic A, Colak I, Jokanovic V, Nikolic N (2008) Matiène æelije zubne pulpe i njihov potencijalni znaèaj u regenerativnoj medicine. Stomatološki glasnik Srbije 55(3):170–179
59. Anderson KE (2009) Huntington's disease and related disorders. Psychiatr Clin North Am 28(1):275–290
60. Hague SM, Klaffke S, Bandmann O (2005) Neurodegenerative disorders: Huntington's disease and Parkinson's disease. J Neurol Neurosurg Psychiatry 76:1058–1063
61. Ross CA, Margolis RL (2001) Huntington's disease. Clin Neurosci 1:142–152
62. McMurray CT (2001) Huntington's disease: new hope for therapeutics. TINS 24:S32–S38
63. Huntington's Disease Collaborative Research Group (1993) A novel gene containing a trinucleotide repeats that is expanded and unstable on Huntington's disease chromosome. Cell 72:971–983

64. Reddy PH, Williams M, Tagle DA (1999) Recent advances in understanding the pathogenesis of Huntington's disease. Trends Neurosci 22(6):248–255

65. Cattaneo E, Rigamonti D, Goffredo D (2001) Loss of normal Huntingtin function: new developments in Huntington's disease research. Trends Neurosci 24(3):182–188

66. Cha JJ (2000) Transcriptional dysregulation in Huntington's disease. Trends Neurosci 23(9):387–392

67. http://www.aboutus.org/Keltner-Inc.com. (Alexa Keltner Inc). Accessed 6 Jun 2012

68. Feng Z, Jin S, Zupnick A, Hoh J, de Stanchina E, Lowe S, Prives C, Levine AJ (2006) p53 tumor suppressor protein regulates the levels of hunting tin gene expression. Oncogene 25:1–7

69. Mitchell I, Cooper AJ, Griffiths MR (1999) The selective vulnerability of striatopallidal neurons. Prog Neurobiol 59:691–719

70. Clement AM, Nguyen MD, Roberts EA et al (2003) Wild-type nonneuronal cells extend survival of SOD1 mutant motor nneurons in ALS mice. Science 302:113–117

71. Beal MF, Hantraye P (2001) Novel therapies in the search for a cure for Huntington's disease. Proc Natl Acad Sci USA 98(1):3–4

72. Jackel RJ, Maragos WF (2000) Neuronal cell death in Huntington's disease: a potential role for dopamine. Trends Neurosci 23:239–245

73. Schilling G, Coonfield ML, Ross CA et al (2001) Coenzyme Q10 and ramacemide hydrochloride ameliorate motor deficits in a Huntington's disease transgenic mouse model. Neurosci Lett 315(3):149–153

74. Rigamonti D, Sipione S, Goffredo D et al (2001) Huntington's neuroprotective activity occurs via inhibition of procaspase-9 processing. J Biol Chem 276:14545–14548

75. Freeman TB, Cicchetti F, Hauser RA et al (2000) Transplanted fetal striatum in Huntington's disease: phenotypic development and lack of pathology. Proc Natl Acad Sci USA 97(25):13877–13882

76. Bachoud-Levi AC, Remy P, Nguyen JP et al (2000) Motor and cognitive improvements in patients with Huntington's disease after neural transplantation. Lancet 356(9246):1975–1979

77. Mazzini L, Fagioli F, Boccaletti R, Mareschi K, Madon E, Oliveri G, Ilaria CO, Pastore R, Huttmann MA, Li CL, Duhrsen U (2003) Bone marrow-derived stem cells and 'plasticity'. Ann Hematol 82(10):599–604

78. Silani S, Cova L, Corbo M, Ciammola A, Polli E (2004) Stem cell therapy for amyotrophic lateral sclerosis. Lancet 364(9429):200–202

79. Mazzini L et al (2003) Stem cell therapy in amyotrophic lateral sclerosis: a methodological approach in humans. Amyotroph Lateral Scler Other Motor Neuron Disord 4:158–161

80. McDonald JW, Xiao-Zhong L, Qu Y, Su L, Mickey SK, Turestsky D, Gottlieb DI, Choi D (1999) Transplanted embryonic stem cells survive, differentiate and promote recovery in the injured rat spinal cord. Nat Med 5(12):1410–1412

81. Sigurjonsson AE, Perreault MC, Egeland T, Glover JC (2005) Adult human hematopoietic stem cells produce neurons efficiently in the regenerating chicken embryo spinal cord. Stem Cells 23(3):392–400

82. Rebuilding the nervous system with stem cells (off the National Institutes of Health Website). http://www.nih.gov/news/stemcell/chapter8.pdf. Accessed 6 Jun 2012

83. Human neuronal progenitor cells. http://www.neuroguide.com/hnpcs.html. Accessed 6 Jun 2012

84. Reynolds BA, Weiss S (1992) Generation of neurons and astrocytes from isolated cells of the adult mammalian central nervous system. Science 255(5052):1707–1710

85. Luskin MB (1993) Restricted proliferation and migration of postnatally generated neurons derived from the forebrain subventricular zone. Neuron 11(1):173–189

86. Kuhn G, Winkler J, Kempermann G, Thal LJ, Gage FH (1997) Epidermal growth factor and fibrolast growth factor-2 have different affects on neural progenitors in the adult Rat brain. J Neurosci 17(15):5820–5829

87. Barnabé-Heider F, Frisén J (2008) Stem cells for spinal cord repair. Stem Cell 3(1):16–24

88. Johansson CB, Momma S, Clarke DL, Risling M, Lendahl U, Frisén J (1999) Identification of a neural stem cell in the adult mammalian central nervous system. Cell 96(1):25–34

89. Huber AB, Ehrengruber MU, Schwab ME, Brösamle C (2001) Adenoviral gene transfer to the injured spinal cord of the adult rat. Eur J Neurosci 12(9):3437–3442
90. Grill R, Gage FH, Murai K, Blesch A, Tuszynski MH (1997) Cellular delivery of neurotrophin-3 promotes corticospinal axonal growth and partial functional recovery after spinal cord injury. J Neurosci 17:5560–5572
91. Cristopher ReeveParalysis Foundation. http://www.christopherreeve.org/research/research-main.cfm. Accessed 6 Jun 2012

Chapter 18
Tissue Engineering and Stem Cells: Summary

If everything seems under control, you're just not going fast enough

Mario Andreti

What are the Newest Bioengineering Approaches that Made a Breakthrough?

The newest bioengineering approaches that have made a breakthrough would have to be tissue engineering (TE) oriented: the use of 3D architecture to create porous, load-bearing scaffolds for bone tissue engineering [1–3]. This creation of bone structure can still allow for nutrient flow and waste removal to keep the surrounding, living bone, viable, but can increase the structure and rigidity of the bone. The use of rapid prototyping and 3D printing is especially interesting because one can create geometries that are impossible with typical machining.

What is Fundamental Requirement for Biomaterials Used in TE and Why?

The fundamental requirement of biomaterials is compatibility. The material being placed in the body must be very carefully constructed, and have compatible geometries, chemical properties, material properties, microstructure, etc. so that it can interact with the host cell's in a desirable way. There is a strong risk of rejection, or a negative reaction by the host, if any of these properties are not appropriate. The material, in general, is meant to interact with the host cells, usually in a positive

M. Pavlovic and B. Balint, *Stem Cells and Tissue Engineering*,
SpringerBriefs in Electrical and Computer Engineering,
DOI 10.1007/978-1-4614-5505-9_18, © The Author(s) 2013

way. However, if the host cell's see the host cells as a foreign body, the host will react by enclosing the foreign body in a dense collagen capsule, and thus, neglect the desired effect of the biomaterial inserted.

What is the Difference Between Artificial Organs and Tissue Engineering Products?

Artificial organs are completely foreign manmade objects, who perform the mechanical or physiological processes of the body in absence of the original organ. These "organs" can include prosthetics, or respirators, or even dialysis machines. These are all considered artificial organs in that they all perform the function of a typical organ, like a kidney in the case of a dialysis, and are completely foreign to the host body. There is a big difference between the development of artificial organs and tissue engineering. Tissue engineering is the process of creating a means to restore function of tissues in the body with minimal rejection from the host. This can mean a wide variety of methods used to produce the desired effect; most of them contain a majority of the host's own cells that have been engineering, regrown, or restored. Typically, if an organ or tissue is failing, it will be engineered by being given the opportunity to grow around a scaffold, or biomaterial that allows for development of healthy tissues [4–6]. This is similar to dying coral reefs that are greatly helped by being given a place to grow around and give it strong anchors. Scaffolds are the same thing, they allow the cells to differentiate and grow, only to regain normal function of the original organ or tissue.

How does CAT accomplish tissue engineering approaches?

CAT accomplishes tissue engineering by the use of 3D printing [6]. The use of this printer involves creating a 3D model using graphics software, which is then built layer by layer using a piezoelectric actuated printer to deposit biocompatible material and create the mold. This is beneficial because the shape and microstructure can be tightly controlled and specified for optimal acceptance and performance within the host's body.

What Does Ink Jet Printer Technique Facilitate?

The ink jet printer facilitates a method of scaffold and cell dispensing at the same time. As of now, the use of these printers has already dispensed human osteoblasts and bovine chondrocytes. Sarcoma cells have been put through the ink jet printer in patterns, and scaffolds created by graphic software. When the cells come out the other side, all three types of cells showed great tolerance of the process. They were all viable, in the desired geometry, and able to grow. This is incredible because new shapes and structures can be created, with real cells, and has the possibility to greatly accelerate the process of tissue engineering, and also increase the accuracy of tissue proliferation, because they can be shaped into desirable forms and geometries. The advance of stem cell research emerges in several directions [7–9], but the main goal is clinical application and their use in regenerative, reparative, and TE purposes.

References

1. Hanson C, Stenevi U et al (2012) Transplantation of human embryonic stem cells onto a partially wounded human cornea in vitro. Acta Ophthalmologica. doi:10.1111/j.1755-3768. 2011.02358.x
2. Giam LR, Mirkin CA et al (2012) Proc Natl Acad Sci. doi:10.1073/pnas.1201086109
3. leah.vardy@imb.a-star.edu.sg
4. Singh AM, Stephen D et al (2012) Signaling network crosstalk in human pluripotent cells: a Smad2/3-regulated switch that controls the balance between self-renewal and differentiation. Cell Stem Cell 10(3):312–326
5. Chhabra A, Mikkola HK et al (2012) Trophoblasts regulate the placental hematopoietic niche through PDGF-B signaling. Dev Cell. doi:10.1016/j.devcel.2011.12.022
6. Kolehmainen K, Willerth SM (2012) Preparation of 3D fibrin scaffolds for stem cell culture applications. J Vis Exp (61):e3641. doi:10.3791/3641
7. Tan TC et al (2012) Telomere maintenance and telomerase activity are differentially regulated in asexual and sexual worms. Proc Natl Acad Sci. doi:10.1073/pnas.1118885109
8. Ishkitiev N, Yaegaki K et al (2012) Hydrogen sulfide increases hepatic differentiation in tooth-pulp stem cells. J Breath Res. 6(1): 017103 doi:10.1088/1752-7155/6/1/017103
9. White YAR, Tilly JL et al (2012) Oocyte formation by mitotically active germ cells purified from ovaries of reproductive-age women. Nat Med 18(3):413–421